Workflow Reengineering

Workflow Reengineering

Gary Poyssick
Steve Hannaford

Adobe Press
Mountain View, California

Library of Congress Catalog No.: 95-81211

ISBN: 1-56830-265-7

10 9 8 7 6 5 4 3 2 First Printing: January 1996

Printed in the United States of America by Shepard Poorman Communications Corporation, Indianapolis, Indiana Published simultaneously in Canada.

Cover and text design by Graham Metcalfe and Paula Shuhert.

Adobe Press books are published and distributed by Macmillan Computer Publishing USA. For individual, educational, corporate, or retail sales accounts, call 1-800-428-5331, or 317-581-3500. For information address Macmillan Computer Publishing USA, 201 West 103rd Street, Indianapolis, IN 46290. The World Wide Web page URL is www.mcp.com.

This book was produced digitally by Macmillan Computer Publishing and manufactured using 100% computer-to-plate technology (filmless process), by Shepard Poorman Communications Corporation, Indianapolis, Indiana.

CONTENTS

Chapter 5 Strategic Approaches: Process Improvement Points 95

Acknowledgments

The authors would like to take this opportunity to thank the many individuals whose patience and willingness to contribute to the project made the work possible.

First, there's Mike Rose. It was during a conversation at Conceppts in Orlando, Florida, that the idea for a book about workflow first arose. We questioned why such a book—one about management and not about software—would be of interest to Adobe® and Adobe Press, the most natural publisher of our ideals. By the time we were finished, many people at the company wound up involved and provided the support, patience, and encouragement we needed to complete the task. Among them were Patrick Ames, editor in chief of Adobe Press, as well as Bob Greene, Eric Bean, Jim King, John Warnock, and Chuck Geschke. We want to thank them for their help. Adobe Systems has provided us with many of the enabling technologies we all take for granted in our everyday work. They know that strategic planning is every bit as important as using the right software and hardware—and this book is proof of that.

Next are two people whose understanding of workflow and business management is well-recognized in our industry: Gary Millet and Minton Brooks. Both have promulgated and supported the basic precepts of this work for years—and the profitability and success of their clients underscore that fact. They both assisted us in organizing our thoughts, and continually provided us with recommendations based on their experience and knowledge.

Most importantly, we would like to extend a sincere thanks to the production and design sites that allowed us to interview them. All of them took considerable time from their busy schedules to provide us with outstanding examples of how to "do it right." They include World Color, Artlab, Gamma One, the *St. Petersburg Times,* Graphics Express, and others.

There are other people as well; Chuck Weger, Paul Beyers, Bob Shaffel; our managing editor, Diane Tapscott; our production and design team of Graham Metcalfe and Paula Shuhert; and a list too long to mention of dedicated graphic arts professionals who gave us advice and direction in our efforts to define the industry's most elusive term: workflow.

—*Gary Poyssick, Steve Hannaford*

Workflow Reengineering

WORKFLOW REENGINEERING

You may not know it yet, but workflow reengineering in the graphics arts industry is bound to affect every individual in the business, from the receptionist at the front desk, to the person passing out paychecks. Why? Because how you reengineer the workflows in your business—whether you are a service bureau, a prepress film house, a corner print shop, a high-flying Web site, a publisher, or a design firm—will make the difference between barely hanging on and achieving profitability you hadn't dreamed of. We know this because we've seen it happen. When we started this book, we thought workflow was going to be a big yawn. "Oh great, an accounting guide for the graphics arts industry." But after months of research in the field, we feel quite differently. This book gives you an insider's view of what we learned, firsthand, from the people whose reengineered workflows have made them successful.

Workflow has become the centerpiece of management rationale in the graphic arts industry today. Yes, quality is important, but many firms have already attained quality

Old vs. New
The graphic arts industry used to have a set of very distinct business categories with relatively clear boundaries. Thanks to the unifying forces of the PostScript® language and desktop publishing, the entire spectrum of graphic arts businesses, from design firms to printers, are being driven together. Prepress and prepublishing (jobs that don't go to press) are now the concerns of almost every graphic arts business.

levels sufficient for their customers. Technical resources matter, it's true, but with the wide variety of excellent hardware and software around, it's not hard to get adequate equipment. Mastering the technology is necessary, but there are many competent, even expert, digital operators available. And certainly astute financial planning and good salesmanship will always contribute to a firm's success. But what distinguishes successful companies from stagnant ones is workflow: its design, its analysis, and its management.

Workflow is neither dry nor marginal; it could well be the lifeblood of your enterprise. It has a major impact on all aspects of your business, from technical purchases, to hiring and firing, to new business opportunities, to estimating, and most centrally to survival and profitability. We have tracked several companies where workflow improvements have catapulted profitability figures from 3% to over 40% in one case. We know that this turnabout was caused by changes in workflow because the personnel changes were few, the new equipment purchases were minimal, and the sales were at the same levels, to essentially the same customers.

It seems you can't pick up a graphic arts journal or conference schedule without seeing "workflow" flaunted about shamelessly. But as hot a term as workflow is, its actual definition seems rather vague. Like many buzzwords, its meaning includes all kinds of functions. And since we don't have a precise definition for the term either, we think its meaning is worth exploring.

What Is Workflow?

Metaphorically, workflow suggests the science of hydraulics and the craft of plumbing. Envision blockages in the system (too many jobs at one station) being cleared by laying more pipes (hiring some freelancers), getting a more powerful pump (upgrading the server), or opening extra valves (buying more RAM). If indeed the workflow in most shops resembles a plumbing system, it's generally a tangle of pipes cobbled together haphazardly over the years and gradually supplemented by newer technologies on a piecemeal basis. Imagine an old house that incorporates iron, copper, and PVC pipes, and combines old and new fixtures that are replaced only when leaks develop or a room gets remodeled—this is the typical graphic arts production house.

You might look at workflow as throughput capacity—similar to wondering how much water can flow through the system. How many jobs, pages, or scans can you handle in one shift, day, or week? Straight throughput is, in part, based on the amount of equipment and power you have. All you have to do to increase capacity is get more equipment. We're certain an equipment salesman would agree with you!

But the truth is, many shops already have more than enough throughput capacity. We know of sites where powerful and expensive drum scanners sit idle twenty hours a day because the machines are so efficient. One shop we know of boasts that its staff could do all the scanning in Southern New Jersey using its super-efficient, high-end drum scanner, given all the time it stands unused. Or take a look at the printing industry—many gleaming eight-color presses run less than eight hours a day, for lack of work to feed them. More important, the combined total of scanners, imagesetters, workstations, and printers in the industry is staggeringly more than is needed. Each individual shop tends to buy more than enough equipment, hoping that sheer throughput power will overcome any problems that might arise. This approach might help meet schedules, but it has a depressing effect on profitability. It also tempts suicidal shop owners to quote jobs at unprofitable margins "just to keep the equipment busy."

The real problem is that even (consistent) capacity is not a one-dimensional concept. Consider our plumbing metaphor—there are four factors required to get the water through the pipes (or jobs through a plant). One factor is the simple width of the pipes, or the capacity of the system as a whole. Capacity along one segment can be speeded up by adding equipment, such as upgrading to fiber optics, buying a bigger hard disk, putting in more proofing stations, or putting in more (or upgrading existing) workstations. The second factor is the speed, or rate of flow. This might be dependent on CPU speed, or the skills of the employees you have; it can also be affected by software upgrades. The raw speed of RIPs (raster image processors) and imagesetters is a third factor. In some cases, if you can be faster you don't have to be wider and vice versa. The final factor is load balancing or flow balancing. That is, the ability to accommodate alternating floods and trickles. Flood stages require alternative workflows to absorb the excess capacity during crisis time, as well as careful planning. As adequate as a system might be for average flow, it also has to handle, in some way or another, peak demand. (On the other hand, contracting out services during critical periods can be effective too.)

Rebuilding the Workflow

In reality, most workflows are not well thought out. Valves are added here, pumps there, and leaks are plugged as you go. It's rare that a computer network and work environment are created for efficiency from the ground up. More often, the typical arrangement is a patched-together system of hardware and soft-

Rework
Whenever a job,or any part of it, has to be reprocessed, whether because of operator error, user mistake, poor communications, or system failure (such as scratched film) rework occurs. Unless the cost of the reprocessing can be recovered (as for author's alterations), rework represents a loss of profits.

Preflighting
Chuck Weger, a consultant with Elara Systems in Fairfax,Va., coined this term for use in the graphic arts industry. It refers to a structured series of tests performed on a page layout file before sending it to an imagesetter. Good preflight practices can significantly reduce rework.

ware having vague dependencies that aren't fully understood. This is similar to having a plumbing system that would scald someone taking a shower upstairs whenever someone else flushes the toilet downstairs.

But graphic arts workflows are even more complex than plumbing systems. Consider quality control, for instance. Imagine if the water in a plumbing system was inspected at regular intervals; noncompliant water would need to be piped backwards in the system to be cleaned.That's just what happens with rework in a graphics arts shop. Rework makes most orderly workflow diagrams useless, since it is rarely taken into account when workflow diagrams are designed. In some shops, 30–40% of the "water" gets diverted back through the system. Even when only 2–3% is recirculated, the unexpected demand creates chaos and delays. Not only does it slow down the daily process, its effects are cumulative, dragging down the whole system. Far from resembling an orderly, well-planned plumbing system, then, many shop workflows resemble a Rube Goldberg contraption.

Another difference between the plumbing model and a graphics workflow is that it isn't just city water running through those pipes. In the old days, when board-based mechanicals were king, most prepress jobs fit pretty much the same pattern, with a little up-front work. Jobs could be regularized to a great extent and most problems could be anticipated in advance—assuming you had a skilled team who knew how to analyze the mechanical and ask the designer a few key questions. Moreover, nearly all jobs were made of some combination of the same basic elements: type, page geometry, transparencies, and opaque artwork. However, with electronic job entry, even with the best preflighting, fundamentals such as color layers, scaling, and resolution may in fact be careless errors or leftovers from an earlier job. Add to this growing list of basic elements—PostScript® fonts, TrueType fonts, RGB TIFFs, CMYK TIFFs, DCS files, JPEG files, CT files, Photo CD files,

Adobe Illustrator® files, FreeHand files—and you'll quickly see that a graphic arts workflow is far more complex than simply delivering H_2O, either hot or cold.

Multiple Workflows

More often than not, it is best to define multiple workflows rather than one giant workflow. In most sites that we've seen, there are at least a half-dozen distinct workflows in operation; in some cases there are dozens. For example, a shop that has both prepress equipment and some sheetfed capability might need five or six basic workflows. Some jobs come with supplied film and require only printing. Of the jobs that come to the prepress department, some are printed internally, while others are sourced out to other printers with a proof. Some jobs require sophisticated color scanning, some are supplied with digital camera files or Photo CDs. And still others require substantial pickup from older work, or supply film from outside sources (such as magazine ads) to be combined with the digital pages. Each scenario needs its own workflow.

Different clients have different needs and expectations. One needs rapid turnaround and only requires a digital proof. Another client, whose work always requires extensive touchup, has more time but wants random, film-based proofs. Perhaps you even run some signage or poster work off the wide-format digital plotter as well. You get the idea. Jobs must be scheduled differently, estimated differently, and may require specialized equipment or skills. Most of all, their differences make it impossible to anticipate that Job A will flow exactly like Job B. While it's true that all jobs are a little different from each other, most shops have jobs that fit into similar categories with only minor variations. This is the work that can be defined and organized in workflows.

The Assembly Line

Some people interpret workflow as an assembly line much like a manufacturing plant where each worker adds one component or checks one function as the job moves on to completion. Indeed, if graphic arts production were similar to Model T production, this interpretation would be valid—but it is not.

Prepress work is only partially a manufacturing process. Imagine assembling a set of automobiles where every model has custom features: designer upholstery, titanium bumpers, solar-powered engines, or custom paint jobs. In prepress, while not every feature of every job varies, every page that is sent to an imagesetter might have its own set of peculiarities. Unlike automobile assembly, there's no guarantee that the parts will fit together until they are on the assembly line. In addition, many jobs require special handling that can back up everything behind it.

Prepress personnel must also be considered. Typically they are not as interchangeable as factory hands. We have found that prepress sites succeed where the craftspeople take an active interest in making sure a job comes out right. Cultivating the ability to spot potential trouble or to correct design errors before they affect production flows is critical to success. A typical automobile model, where workers are responsible for their own narrow functions, simply doesn't work.

In many ways, we believe that prepress manufacturing has to be standardized as much as possible. But standardization is not an absolute rule. The weakness of the typical assembly line approach is that it is too rigid to adapt to changing conditions or to find the best way to do a job. A shop that converts every PostScript file to Scitex format, whether it makes any sense or not, is not efficient. A traditional assembly line approach is likely to misfire, since it is not flexible enough to handle variation. And variation, while it needs to be managed, is inevitable.

Job Jacket Mentality

OPI
Open Prepress Interface creates a low-resolution file from high-resolution scans. These low-res files are used by the designer to indicate the position of an image on the page. When the final pages are rendered to film, the low-res files are automatically replaced by the hi-res scans.

Another traditional model that helps define workflow in a prepress shop is based on the idea of a job bag or jacket; that is, a collection of pieces belonging to an entire job or some significant part of it (a chapter, signature, or spread). The job jacket carries the job through the plant while various elements (type, line art, images, tints) are integrated. While some parallel processing occurs, basically the whole job marches through the shop together. Traditional workflow management solutions are based on making that march run as smoothly and efficiently as possible.

Despite the digital revolution, that old job jacket mentality still persists. There may be some OPI (Open Prepress Interface) used, there may be an optical disc or a few SyQuest cartridges in the job jacket, there may be better use of a central server, but in general, the old workflow rules still apply—even though the method of working has changed radically and deadlines are tighter than ever. Many shops still have "job jackets," but now they are digital.

Air Traffic Control

To better understand prepress workflows, think instead of an airport (we have expanded an idea from Barry Rickert, head of Digital Color Solutions). Workflow management resembles the job of air traffic controllers, who must try to keep landings and takeoffs on schedule while safely allowing planes to come and go. Inevitably, in busy airports, no matter how good the schedules or the traffic controllers, failures occur and traffic gets backed up. Once this happens, the controllers scramble to keep traffic moving as airplanes wait idle on the ground or in the air for available gates or runways. The slightest mechanical breakdown can paralyze portions of the operation, and worse still, a whole set of interdependent functions—connecting flights, baggage transfer, staff coordination—can be seriously affected by a major event (such as a snowstorm).

In the same way, prepress shops must be prepared for emergencies and mechanical failures. Clients constantly bring files in late or with the wrong elements. A "down" imagesetter or workstation can cause serious routing problems. Jobs arrive early, or even on time, only to sit stacked up waiting for another job to be finished. Schedules start to slip, transparencies get misplaced or sent to the wrong place, fonts get lost, and customers begin to panic. Even worse, unlike the airport, the prepress shop is acting with almost no radar; schedules change daily, even hourly, and jobs can arrive haphazardly, without warning. No wonder so many people in prepress departments spend hundreds of hours at scheduling meetings—a necessary task, but one that doesn't generate any profit.

A FINAL METAPHOR. Think of the sidewalks of Chicago or Manhattan at lunchtime on a pleasant day. On any major street there could be thousands of people walking from one place to another, each with their own errands to deal with. Yet amidst the seeming chaos, and despite the different rates of speed the people are walking, collisions almost never occur. People unconsciously adjust their steps to flow with each other. Whether others stop, slow down, or speed up, people manage to swerve and sway, seemingly without thinking. Everyone gets where they are going, more or less on time. Minor mishaps or congestion rarely stops anyone for more than a few seconds. The system requires little supervision—pedestrian lights at the intersections—and yet is extremely efficient. We think that a well-managed workflow should run as smoothly. It does not require constant attention from managers, schedulers, and CSRs. If routes and destinations are well described, and excess work does not get out of hand, a workflow should run quietly along despite interference from all sides.

Systems Analysis

A key tool for success is a good financial analysis of your business. Analysis starts in the accounting department. A clear understanding of profit and loss, cash flow, monthly equipment costs, and current profit ratios are important to any business, especially a service-oriented one. Executives that don't review these figures from week to week, or at the very least, month to month, are sailing through the shallows without a chart and are likely to run aground.

But successful companies also need a more detailed level of information so that they can compare accounting numbers to production numbers. Which customers are making you money, and which are making you lose money? How accurate is your estimating? Which equipment is profitable, and which is losing or merely breaking even? What do you base your pricing on? Which personnel are good performers, and which are not?

On still a deeper level, you'll want to know about the details of your workflow. How many pages are processed a day? What jobs get routed through which stations? How much rework is being done? Where and why do schedules slip? What kinds of activities have to wait until one or two key people get around to it? How many hours of nonbillable overtime is performed?

And finally, you need to make month-to-month comparisons: are we getting better at any new processes? Did the changes (in procedures, equipment, personnel) we made last month make a difference in this month's figures? Is the direction of overall profitability improving?

Part of this book is about organizing data (aside from the pure accounting data) to get the answers to these questions. But before you can obtain data and put it into a report, you must first understand how the actual workflows in your plant work.

Back to Front

The sanest way to look at workflow is from back to front. Ideally you want to produce a stream of profitable, acceptable work from the final output devices in the workflow, whether printing presses, imagesetters, or copiers—work that can be confidently shipped and invoiced. The duty of the rest of the plant is to keep that flow humming steadily. The further downstream—that is, the closer to the output device—that errors are caught, the more expensive and disruptive they are to fix. Thus, it makes the most sense to correct problems upstream, even, if possible, when they are still in the designer's computer. Most estimating and scheduling are based on machines and employees who do each job only once or (as with color corrections) a predictable number of times depending on the pickiness of the customer. Anything else is a disruption of workflow.

In this book we propose methods for understanding and diagramming actual workflows. It requires the considerable task of breaking down processes into individual steps. You will probably find that certain sequences are common to many workflows. Diagramming or mapping workflows is serious and slow work, but once done it gives you a formidable tool for clarifying, measuring, and redirecting the business you do. As a general management tool, a workflow diagram can help pinpoint serious problems in the way certain jobs are performed. For sales personnel and estimators, it helps pinpoint exact turnaround times, both for familiar and new workflows. For technology managers, it can highlight situations where new technology might make a major return. For human resources, it can indicate areas where training or cross-training would improve performance. For accounting, it helps assign job costs. For personnel in general, it can clarify areas for overall improvement. And it also gives a basis for predicting the human costs and potential ROI (return on investment) in adopting a new technology, such as digital printing.

Workflow and Profit

The main reason to improve workflow is to enhance profitability. An increase in efficiency means nothing unless it is tied to such enhancement. Even increased customer satisfaction, the purported goal of many service industries, is useless if you satisfy your clients and still go out of business.

Workflow enhances profitability by making you more productive. If you can improve specific workflows within your organization, you'll see your profits rise. For instance, minimizing rework is worthwhile because it has a direct impact on profit. Identifying the workflows that are profitable and those that are not can help focus sales efforts. Tracking equipment workflow will help you decide if future purchases will actually improve your bottom line. Inefficient processes can be analyzed and revamped depending on profitability. And all employees in the company, especially if management structures things right, can begin to think of improving profitability by improving workflow.

People

In most shops, equipment is only part of the BETTER WORKFLOW = BETTER PROFITABILITY equation. Personnel costs make up between 60–70% of the budget in most graphic arts shops. It seems that many business executives spend too much time talking to salespeople and attending trade shows, trying to buy that one magic piece of equipment that will change their lives, while the real solutions lie in personnel. Finding ways to make your people work more efficiently is perhaps the most important workflow solution, and while sometimes this means upgrading hardware to help them be more productive, more often it means improving the way they work.

This is not a punitive process. Most people want to work well and efficiently. Once employees understand that their job, and perhaps part of their remuneration, depends on overall profitability, and therefore the efficiency of the shop, most employees become interested in improving workflow. Unfortunately, most people in our business aren't given the time, nor the tools, to improve their own performance. In our experience we have found that employers cannot simply demand better performance; rather, they need to offer tangible incentives if they expect results. Motivating people to take responsibility for the company's mission should be one of the main goals of management.

The Art and Attitude of Management

TQM

We have seen Total Quality Management in practice at several prepress and printing companies. We applaud the good intentions, but there seems to be a mixed record of success. It's not that the statistical controls fall short, but rather the ability to hand over responsibility and power from management to workers.

Whatever the latest management buzzwords, many managers secretly believe that the only way to get the ship up to speed is with a cat-o-nine-tails and by keelhauling the worst lubbers in the crew. For the more enlightened, other management philosophies exist: TQM (Total Quality Management), managing by walking around, and empowerment, for instance. Developing a management style that suits your company's personality and its profit goals, clients, personnel, and workflows will probably yield the best results. We suspect that many of this book's readers feel they are not at a high enough level to effect change. Don't believe it. The power of a good idea—one that produces a good workflow—will be listened to by any profit-motivated manager.

By the way, much of what we propose is consistent with

TQM. But we are not doctrinaire Demingites—the custom manufacturing nature of prepress does not exactly fit the typical TQM model. We also think it's difficult to impose doctrines from above, and that TQM often fails because employees feel free to ignore yet another management fad. The workflow-oriented approaches we describe end up with TQM-like results, but they are geared toward the graphics arts industry.

An empowering management style does work—we've seen it again and again. Employees who take responsibility for their work generally help produce higher profits than if they don't. One way to empower employees is to keep them informed.

For many owners, the thought of disclosing their company's financial position to employees is even scarier than offering them profit sharing. We are convinced, however, that sharing financial information with your staff is a very smart move. After all, it's hard to be a team player when you don't even know the score—not to mention that jobs and paychecks depend on the profits of the company. If profits improve, jobs are secure, and paychecks could possibly increase.

A large part of improving workflow means improving management. Make it possible for employees to succeed and they will work hard to do it. This does not mean abdicating leadership—far from it. Nor is it a matter of sloganeering or buzzwords; experienced employees see right through that. We think defining clear objectives that employees can share will impact your profitability in a very positive way.

Workflow reengineering, by the way, is not as easy as writing a purchase order for a new UNIX-based server, or rolling the dice on a digital printer. While we are a bit glib, we are well aware of the difficulty of adjusting to a truly workflow-oriented management structure. Corporate culture, personality problems, and the "this is the way we've done it for thirty years" mentality are among the obstacles to overcome. Chances are, you'll ruffle some feathers if you end up adjusting sales compensation, or redefining middle management, or try to balance status and perks. You may even have to change your own behavior. But, from what we have seen in real life, it's the only route to continued profitability in this industry. Look at it this way—you could end up with a less stressful, more satisfying, more profitable business to run.

CHAPTER 2

FROM CREATOR TO PRODUCER

Workflow isn't confined to the manufacturing environment. Workflow starts the moment a designer is assigned the task of massaging information into a meaningful, effective container—that is, the output medium. The medium might be a printed brochure, an annual report, a book, a newspaper, a compact disc, a video, or a package. The output medium for the message determines the workflows required to create and deliver the communication. As we have found, there isn't one workflow that describes the perfect path a project should take; instead, there are many smaller, individual processes that combine to make any number of possible workflows. As a producer, you determine which processes, or workflows, are needed, chain them together, and hopefully make money carrying them out.

The Old Workflows

Desktop technology was first adopted by creators. Before that, standard workflows were well understood by everyone involved. The designer or publisher was responsible for the creative tasks and the producer brought the job to fruition using the mechanicals provided by the creator (or client).

Each side of the creator/producer equation had well-defined limits. Few designers, for example, knew what choking or spreading was, much less assumed responsibility for the accuracy of that process. On the other hand, there weren't many color separators who also had designers working for them. They felt, justifiably perhaps, that offering design services made them competitors of the very companies that made up their client base. It was the same with printing: although a prepress house may have understood printing, they didn't want to alienate the printing market; it represented their life's blood. Designers designed, manufacturers manufactured. The lines were clearly drawn (in most cases), and everyone knew exactly what their role in the process was. This compartmentalization worked until the introduction of the first Macintosh.

Traps
In discussing traps and trapping, we mean what some call "chokes and spreads," which are not to be confused with ink trapping. Digital tools have made a major impact on trapping in the trade. TrapWise and Scitex's FullAutoFrame support page layout practices that manual methods could never accomplish.

There was even a certain degree of mystery about what other people in the workflow did. Separators didn't really understand type selection, object placement, trim sizes, shapes; they didn't care how readers interacted with the page, or about the sizzle of a good headline. They didn't need to. Their role didn't require a solid understanding of the design workflow. Data moved from the design site to the production site in the form of written instructions on a tissue overlay. Everything was visible and pretty much standardized. Traps, for example, weren't indicated on tissues—but color breaks were. The designer picked the colors; the separator dealt with the mechanics of overlapping or adjoining objects during film production. Each had a job to do, and they knew how to do it. The designer didn't care if the separator was sprinkling goat's blood on the films, as long as the printed piece came out as designed and specified on a tissue overlay. Designers didn't notice traps, because traps usually worked. They didn't care about GCR or UCR, or even know what the terms meant. Similarly, printers and separators didn't concern themselves with kerning or tracking tables—the type was already in place when the materials arrived. Type was a line shot; it never stopped the workflow dead in its tracks.

The industry enjoyed a relatively predictable timeline. True, tight deadlines were always a reality. But since everything could be seen and anticipated within existing workflows, you rarely found any major surprises hidden behind, inside, or within the tissue overlays. Problems like missing fonts within Encapsulated PostScript (EPS) files just weren't there. Creators and producers knew exactly what to expect, and how much time various processes would consume. In a crisis, you could throw more strippers at the job. You didn't RIP anything—you shot it on a camera, burned it on a frame, or stripped it on a table. You used razor blades, tape, and films. Workflows were largely quantifiable and could be scheduled accordingly. And, most important, you knew which workflows were profitable and which ones weren't. You didn't attempt to do Flexo plating if your specialty was sheetfed annual reports—or vice versa.

Who Is the Creator?

Creators represent a wide range of individuals and organizations involved in the inception and design of communication vehicles. In this book, we address commercial creative environments—not the customer who needs thirty copies of a church flier.

While desktop technology has put pressure on the producer environment, it has helped to grow the creator environment. In short, the client base is expanding. The challenge for the creator is to stay in business while catching up technologically—constantly having to evaluate the feasibility and effectiveness of new media as they arise. Like the producer, the creator needs to develop new workflows.

Taming the Wild Workflow

Despite our claim that front-end workflows have changed radically—and they have—there is still an identifiable series of steps that a project goes through before it can be shipped. The more you know about each stage of the process, the better prepared you'll be to establish profitable workflows within the context of your business.

The key to analyzing and taming workflows is simplification. While no graphic arts producer can or should force every job through the same set of processes, imagine the chaos if you had to invent a new workflow for every job. Even a new job type should have some processes in common with familiar or existing workflows. When no commonalty exists, it is impossible to make money. Taking on a brand-new, unfamiliar, and unplanned-for job is as unlikely to be profitable as delivering pizzas in your spare time. For that reason, many shops have learned to firmly refuse that stray job laid out in WordPerfect or Microsoft Publisher for the PC, for example, and send the client to those few shops that happen to specialize in such jobs.

Job Definitions

Once upon a time, you knew how to define everyone's job. Creators compiled the content (copy, illustrations, layouts, photographs) and partially assembled these components into a physical mechanical. Then they took their mechanicals to the producers, who processed the components. Upon completion, the vehicle was distributed to the audience.

The producer side of the workflow consisted of several specialized talents. There were typesetters who composed text elements; prepress experts who input and output photographic elements and separated them into CMYK for processing on press; camera people who shot line art into appropriate sizes; and printers who got the entire project on press and produced enough impressions to meet the project's requirements. In various combinations these disparate groups worked together—each controlling their own workflows. There was very little interaction between the groups beyond solving specific problems or seeking answers to specific questions. Each step was measurable in terms of time and cost: from setting good type, to producing quality halftones, to printing great 150-line screen color pages, there were required skills for each step. Desktop has essentially destroyed this well-defined process map.

Color Gamut
Color gamut is the range of colors that can be reproduced by any device. Many colors that the eye can see cannot be reproduced through four-color process printing. With the wide variety of color output devices and proofing tools currently available, along with new strategies for printing, such as Pantone's Hexachrome system, a six-color process printing, the issue of reproducing gamut is increasingly important.

In predesktop workflows, what could and could not be done was, for the most part, known. There were specific things that the creator could do to slow down or stop the workflow, but they were somewhat limited. They could, for example, issue instructions to a typesetter that couldn't be executed (set *War and Peace* in this quarter-page ad, please). Or they might select color combinations that were impossible to produce with process colorants—forcing the job to run with a fifth, or even a sixth, color to reach their creative vision. They might specify tints that were out of gamut, or create artwork with dyes that a scanner wouldn't see. They might have created complicated vignettes or outlines that could be done, but were very expensive. In general, though, creators were very limited in their ability to stop a job, and producers often had workarounds they could apply with minor impact on schedule or price. The most effective creators talked with their vendors before designing jobs that required skills beyond what was well known. The more experienced the designers were, the less likely they were to design projects that couldn't be manufactured. And since most printers and separators worked within the same constraints, their "systems"—or workflows—were all more or less identical. That's no longer true.

Job Killers

Nested Graphics
When you place an Illustrator EPS file inside a Photoshop® TIFF file, and then inside a FreeHand EPS file, for example, you have created a nested graphic that will most likely give you trouble when you print. This is because the possibility of bad fonts, improperly specified colors, and similar problems existing in the nested file is very high.

Today there are many problems—far more than before—that might stop a job, or push its required processing time outside an acceptable range.

Examples abound: knockouts and overprinting often cause problems because it is too easy to assign incorrect attributes; gradients can be designed with color combinations likely to cause stepping or moiré patterns; gradients can also be designed with too few, or too many, steps between colors; kerning tables get lost and clients' font versions don't match those found on the producer's system; Encapsulated PostScript artwork containing placed images or nested graphics wreak havoc at printing time; and application version incompatibilities crop up constantly. The list goes on and on.

If producers advertise "We take desktop files," they have no idea what might walk in the door. This problem strikes at the very heart of profitability. If you can't quantify and anticipate any possible permutation of a mechanical, you can't develop profitable workflows. We all know of a job that took six hours, or all weekend, to RIP, and still resulted in bad film.

Another thing to consider when designing profitable workflows: creators who decide they need control over processes that were once purely the producer's responsibility. Some designers, armed with how-to books, scanners, and various output devices, execute portions of their mechanicals incorrectly and then expect you to fix what they messed up. How do you do that?

As we mentioned in the opening chapter, this book focuses on the ink-on-paper communications model. Now that the

World Wide Web, CD-ROM publishing, and digital printing are viable options, establishing effective workflows for these media is also vital. The ink-on-paper workflow model isn't limited to output for printing presses either—the quality controls you need are even more important in these newer communication, vehicles. Consider, for example, direct-to-plate or digital printing. If your current workflows result in unacceptable wastage and spoilage, imagine the costs you'll incur when trying to image plates that can't be fixed with a razor blade or by simply pulling a new single-page form.

Workflow and Training Upstream

Device Profiling
Device profiling creates a file that documents the color characteristics of an individual device, such as a monitor, a scanner, or a color printer. These files can be used to filter color information as closely as possible to a standard color representation.

The only way to control how creators prepare their files is through a structured method of upstream training. That means training for the creators who initiate projects from the start. Is training part of your workflow strategy? Is it part of your budget? If not, there's no way to control the many variables that creators can introduce to the mix. Do your clients do their own color correction and then expect you to match what they see on their monitors? Color calibration aside, there's still a solid argument that states that unless you do color "by the numbers," there's a good chance something will go wrong. The ColorSync color management system and device profiling might offer great strides toward portable color, but accurate analysis of an original image, for instance, isn't something that comes with Adobe® Photoshop® or your scanner software. Producers have skills that many clients need to learn, or at least understand, if they're going to remain profitable clients.

The graphic arts landscape and the animals that roam it are changing. The client definition you had in mind five years ago won't work tomorrow or next year. You need to be in close contact with your clients. You can no longer rely on them doing their thing and you doing yours. It just doesn't work.

ColorSync
ColorSync is Apple's programming interface that lets developers use device profiles. The profiles are supplied by the device manufacturers or by color management programs. ColorSync enables them to be used to make color adjustments that ensure greater color accuracy.

There is incredible pressure on the creator to push the envelope of technology. The very same manufacturers that sell you your imagesetters are knocking on their doors too. They have entire marketing programs aimed at telling your clients that they really don't need all your services anymore. While generally not true, it sounds attractive to the creator: "Gain more control! Save money! Simply buy our ColorPixRipStation and your problems will all go away."

You have to be the ultimate expert in the minds of your client base. Gain their trust and they'll seek your advice. If they can profit from doing some of their own scans, teach them how to do it. Heck, you might even buy them a scanner and put one of your people on site to help them out. It's better than losing them to a competitor who might use these sorts of tactics against you. Remember, clients are related to workflow and profitability. You need to understand them to control them. Make it easy to do business with your company. Profitable and effective workflows are what you sell—not time on your presses or your ability to do a good scan.

External Responsibility

How your business worked ten years ago doesn't amount to a hill of bean sprouts today. Your clients have changed. They have powerful tools capable of doing all sorts of (previously) production-specific tasks. If you fail to define what they should and shouldn't be responsible for, you can bet that they'll do things you'll have to undo. And undoing things is neither productive nor profitable (nor will it win any applause from your client). You can help your clients by defining workflows that take advantage of their design skills and tools and leave the production-specific tasks to your staff and equipment. In other words, define workflows from start to finish, and find cooperative clients. There are some that won't want to work under those conditions. If so, you might need to eliminate them from your sales forecasts because you won't ever make money trying to fix something that they continue to do incorrectly.

To maintain control of your business, clearly define who—you or your client—is responsible for which processes and workflows. Without defined expectations, the contract between you and your client becomes ambiguous. Who, for example, is responsible for color correction? And then there is trapping, clearly a function that is press- and production-specific. Does your client make her own traps? Which tasks should you do, and which should the client do? Who pays for the creator who insists on assuming responsibility for production-specific workflows? You do. Clients want their lives to be made easier by their service providers—not more difficult. If your employees throw their hands in the air and blame clients for not knowing production-specific processes, who is to blame? The client? Wrong. That attitude will get you to the poorhouse in short order. A service provider who uses blame tactics doesn't have control over his workflows.

Internal Responsibility

Does your site still rely on dozens of departments separated by culture and floor space? Do you have a customer service department that doesn't understand the meaning of preflighting, or isn't equipped to perform this vital task? Do you have a prep department that's separate from (and at war with) your color department? Do your strippers know what they're seeing when supplied with reader spreads generated from a client's imagesetter? Look hard at departments that worked perfectly in 1984 but continually fail to effectively move work in today's climate. We see it daily at sites everywhere in the country. Examine your organizational structure closely, looking for places to minimize duplication of effort, to expand or strengthen communication, and make appropriate adjustments based on the business you're doing today and hope to be doing tomorrow.

Working with Clients

Companies that provide services to an increasingly computer-savvy client base need to define services that fit the needs of that market, while developing a much closer relationship with them than ever before. Simply bringing in a pile of the latest equipment, with the hope that it will solve your workflow problems and those of your clients, is not a real solution.

Your clients are having as hard a time with their vendors (you and your competitors) as you are having with them. It's a two-way street. Since they're not manufacturers, and the solutions to these problems are manufacturing-related (production-specific), you're the person who has to provide the solution. It's not going to come from the creative community. Any such solutions (buying their own imagesetters, scanners, or expertise) are not likely to include you in their business. Anticipate where your clients' needs are headed and get there before them. If not, you'll be left in the dust.

Clients want to sleep at night. Pulling their own film when they are supposed to be in bed all cozy and warm is not their idea of a good time. Most want you to control their color from scan to press so that they can get back to what they like doing best—design. If they do want to do their own front covers or artistic retouches, teach them what they need to know. As soon as they realize what it means to correct 120 images for one brochure, they'll be at your front door, we can assure you.

Clients who insist you're too expensive or too slow because they know how to use a page layout program and Adobe Photoshop aren't ever going to feel differently—unless you give them a good reason to. Fix your workflows—develop smooth, profitable, and effective ways to process their work, or provide them with whatever they need, and they'll love working with you.

Of course there will always be the "ego" designers who think they've become Mr. Retouch or Ms. Correction—but they're not in the majority. Sooner or later their employers will realize what a resource drain this approach produces within their own creative workflows. Although it seems the line between creator and producer seems to be getting muddy—it's not. The creator is responsible for the layout and rhythm of the electronic page; the producer is responsible for making those documents sing when it's time to run film and print and ship the job.

Specific Problem Areas

The new desktop tools your clients are using have brought a unique set of workflow-related problems to the production floor. Each component of a page represents a potentially flow-threatening process. Work with your staff to identify areas where problems often occur. (See page 122, "Catching Output Errors.")

COMPLEX ART. In the old days, conventional mechanicals rarely contained blends trapping against blends; four-color type (it was usually provided as a line shot and you kept it black); images inserted within type outlines; or illustrations that contained links to other illustrations or photographs. While the creative possibilities seem boundless, keeping up with design in today's desktop environment can be a production nightmare.

MIXED SCANS. Another potential problem that can slow down your workflow is that all the continuous-tone images, or even parts of a composite image, on a single page might not have been scanned on the same scanner. This can easily create a situation where correcting one problem causes another—

adding cyan to whiten one image, for example, might make another image cast purple.

Color Management Vendors have promised the world with a host of color management products aimed at calibrating devices and characterizing output. But despite the hoopla, desktop color management has yet to be realized. A new generation of color matching profiles, along with Apple's second version of ColorSync, shows promise, but only promise.

COLOR MANAGEMENT MISCONCEPTIONS. Who is responsible for such color problems? Clients have been told that color management systems will solve these fundamental color problems. That's simply not true. The type of problems that arise when images are scanned on different scanners, for example, is particularly troublesome when the designer composites two or more of these images together. Were the white and neutral values balanced in the original images, or after the images were merged? Who is responsible for ensuring that white stays white—the way it looked on the client's monitor? This is an area where promises made by manufacturers of color management devices are sure to cause problems for you, unless you anticipate them and adjust your clients' expectations. Just because two devices are calibrated doesn't mean that suddenly your clients will be trained to distinguish an image carrying a cast from one that is truly neutral.

TOO MANY OUTPUT OPTIONS. Workflows once had a limited set of output options. In many cases you generated (or bought) random color whose input and output drums resided in the same box. These were assembled using a standard set of line screens—shot on a camera or assembled on output devices from a single vendor. Not anymore. You never know where supplied film might come from. Chances are, you own two or more output devices from different manufacturers. These types of potential problems can be minimized or eliminated by defining workflows that anticipate such situations.

CLIENT APPROVAL. Proofing is changing. Clients still expect high-quality proofs, but their definition of what's acceptable is evolving. Some refuse anything but conventional film-based, layered proofs such as Cromalin™ or Matchprint™. Others want

Digital Proofing
The main areas where digital proofing is growing is in direct-to-plate and direct-to-press systems, as well as for nonhalftone screening processes. In cases where there is no film, film-based composite proofs are impractical. For situations such as stochastic screening, digital proofers can simulate nonhalftone-based dot patterns better than film-based proofs can.

digital proofs. Since the price of digital proofing systems is definitely heading down, you might consider one. (By the way, digital proofing is an area where color management systems are proving to be worthwhile by ensuring realistic and consistent press simulations when outputting digital files.)

PROCESSING PITFALLS. Increasing the speed at which you can process a client's page might give you a false sense of security. Have you ever found yourself explaining to a client why—after three days—you are only now calling to get a missing font or EPS file? You thought you had plenty of time to produce the job, so let it sit unopened. They felt the job should be on press already and you're just opening the disk now? This scenario strongly supports the importance of timely and structured preflighting, as well as making preflighting the first step before all other processes in your workflows.

Who is the Producer?

What is a producer today? Many have expanded their services far beyond what they might have felt comfortable with a few short years ago. Training clients makes sense and might make some money in the process, for instance. So might providing design services. Many potential clients don't have extensive design capacity. While a publisher or advertising agency might employ dozens of designers, a manufacturer of bicycles might not. They need brochures, annual reports, point-of-sale materials, and many other marketing-related materials, but where do they go for design? A local agency who then marks up your services before they sell them to the client? Why not provide design services yourself? Are you afraid of upsetting your agency clients? While this might be a possibility, it will hurt your business far less than you think. Some clients might revolt, but very few will see you as a threat to their core business.

The option of offering design services isn't just our opinion; many top printers and separators have added print and interactive design to the services they provide, with great success. Apart from doing straight design for some of your clients, your in-house design team can also act as backup support for other clients during critical periods. And sometimes, having a designer on staff might help you communicate better with the clients' designers.

Another area in which your company could expand is providing new input sources. Kodak Photo CD is, for example, a very popular method for low-cost image capture. Designers are flocking to the technology in droves. Before you jump in, though, there are several things to consider.

First, the raw images can have terrible tonal balance. You'll need to offer image correction services (and keep the price in line with the inexpensive nature of the original media) or teach your clients how to at least set highlight and shadow dots. We're not talking about critical color correction here, just proper distribution of the existing tonal values across the range of available grays.

Color Separations in Adobe Photoshop
Many people make excellent separations in Photoshop, especially with the plug-in tools now available from a variety of vendors. But it is also possible for people to make bad decisions about color and separations, especially if they are unaware of their printing requirements. Making good separations still requires expertise and experience.

Second, do you have effective workflows for handling RGB images? Or RGB JPEG images? Or, as is the case with many service providers, do you simply deal with stuff as it comes in the door? Converting RGB to CMYK was never a problem when you had a big, bad color computer hidden in the dark recesses of your $250,000 scanner. Now, though, many clients are separating images using Adobe Photoshop. While Photoshop can produce good color for a variety of projects, it still takes a certain level of expertise to produce good separations—a level your clients may not have. If your clients want to work in Photoshop, you might encourage them to leave the files in RGB so that you can manage the conversion to CMYK.

Alternative screening options are becoming more prevalent. Stochastic screening, for one, has made headway, and it is being used in some niche markets such as top-of-the-line automobile brochures. If you want to take advantage of these new screening technologies, make sure to develop workflows specific to the generation, proofing, and output of these random dots. Clearly, you can't simply slip these radically different separations and films through a normal workflow that is based on a conventional dot. As of this writing, we've seen only a limited number of companies succeed with this technology. But it does provide promise for the future.

If you are working with stochastic screening, creating a profitable workflow to handle these projects is vital. Digital printing and direct-to-plate are other new technologies for output that require their own well-defined workflows. The point is, any new work models will require a new workflow to ensure profitability.

In summary, desktop creative abilities are forcing producer's workflow models to strain and crack. Too often, workflows, once established, tend to last for a couple of years, maybe less. A couple of years in the ever-changing digital domain can be deadly.

Creators have a broad range of creative possibilities that can break the back of even the most well-thought-out workflow. Depending on how far upstream you're willing to go, producers need to venture beyond a purely downstream environment.

CHAPTER 3

Real-World Operations and Their Workflows

Ultimately, any discussion about workflow is only as good as its applicability. How does a reengineered workflow fit in the real world? What can (and can't) be done to improve the graphic arts workflow where there are computers on everyone's desk? We decided that the only way to answer this question was to go out and see for ourselves.

We selected a variety of graphic arts sites: from designers to printers and everything in between. These are, of course, not the only experts out there. There are hundreds of shops busily engaged in creating or changing their existing workflows to improve productivity, reduce rework, and increase profits—and many are succeeding.

We have chosen sites we think are exemplary. Each has a uniquely designed, well-executed approach to solving productivity problems. We selected a range of projects and approaches that spanned different business areas and delivery systems, such as monitors, digital color presses, the World Wide Web, and large-scale web presses.

Most important, individuals at the sites were willing to spend the time to be interviewed. As is the case for most of us, time is a precious commodity, and these people gave freely of theirs in an effort to help us write this book. We take this opportunity to thank them one more time. You may recognize some of their names as well, from industry conferences, magazine articles, or from your own research efforts.

CASE STUDY

LANMAN LITHOTECH

Custom Workflows

The Lanman Companies were founded in 1911 in metropolitan Washington, D.C. In the eighty-odd years since the company started providing zinc-plate etchings for the printing industry, they have grown to become one of the nation's leading suppliers of high-quality graphic services. Their work with clients such as Smithsonian, National Geographic, Publisher's Clearinghouse, L.L. Bean, Wearguard, and many others have, over the years, established their work as a barometer against which the quality of high-end color is measured. The company remained family-owned for four generations. Very recently, the company was purchased by World Color, making the parent corporation, at well over $1 billion in annual sales, one of the world's preeminent production environments.

Bruce Cunningham, president of the Lanman Lithotech division in Orlando, Florida, has long been recognized as an innovative leader in the field of digital page construction. Under his direction, the company's imaging operations have evolved from a Hell/Scitex proprietary environment into one of the most respected Macintosh-based digital prepress houses in the world. With no reduction in quality, Cunningham successfully migrated a huge, well-run prepress operation to a fully open-architecture environment, while many of his peers were still wrestling with the basic concepts of outputting PostScript files and merging them with high-resolution images. There is enough material within the Lanman organizations to fill an entire book about digital page construction and required workflows. For example, in Chapter 6, "The Database," we relied on Bob Nuelle, Director of Advanced Technologies for the firm, to provide much of the technical guidance. Some of the many approaches the company has taken to digital workflows include: designing workflows that match their clients needs, using databases to manage large color inventories, and using the Adobe PDF (Portable Document Format) file model as a remote proofing scheme.

Client Strategy

Lanman's typical client profile is one that contracts for long-term, repeat work. Workflows are usually designed and hammered out long before the first pages actually arrive for production scheduling. Fonts are agreed upon and purchased by Lanman if they don't already have them. Lanman representatives are made aware of rough image inventory early in the design process, and the company supplies design services to the client during peak periods of activity. Such long-range, careful planning is possible when work comes in on a relatively well-structured schedule. Predetermining workflows for specific clients and designing them in advance is the cornerstone of Lanman's corporatewide strategy for providing high-level service to its clients. This approach involves a broad cooperative effort that includes many representatives from disciplines on both sides of the creator/producer equation. Lanman feels that maintaining such an efficient service business requires the highest standards, and that the only way to control rework, profits, and satisfaction levels is through proper workflow planning. The company relies heavily in all of its divisions on strategic planning, and this effort extends deep into all of their relationships. Lanman's management and staff universally agree that the more you're involved in every step of the workflow, the better you will be able to control production variables. Only through proper analysis and training (another major element in the Lanman strategy) can you reduce rework.

File Strategy

Most of Lanman's clients create rough page comps in Quark XPress on Macintosh workstations. Lanman routinely provides its clients with the Adobe Prepress XTension, an extension that, when used with XPress, intercepts Quark's interpretation of the PostScript code and parses it to bring it closer to the published Document Structuring Convention (DSC), which is a

DSC
Document Structuring Convention (DSC) is a set of rules and standards that applications should use to structure PostScript and EPS files so that they can be processed predictably by other applications.

fundamental component of Adobe's PostScript Level 2 strategy. This specification defines the structuring conventions within PostScript output files. Since XPress's use of PostScript doesn't conform strictly to the guidelines of DSC, the XTension helps standardize the incoming PostScript files at Lanman's sites—most of which are in the form of XPress files when they arrive. The Adobe Prepress XTension operates transparently within Quark XPress and is available at no charge from most of the major on-line services; you can also download it from the Web.

When files arrive at Lanman, they are typically in native (application) format, including added disk files containing any high-resolution images that the client may have worked on directly as a part of the creative process. The majority of the images have normally been scanned at Lanman before the arrival of the final pages from design, and have been placed at the client site in low-resolution OPI format. These images are automatically replaced with the high-resolution images at Lanman when pages are output.

Basic Lanman Workflow

1. Early in the design process, immediately after the creation of rough page designs, images are selected and transparencies are sent to Lanman.
2. Lanman executes high-resolution scans, retouching, and color correction based on their own analysis and client specification. At the same time, the designers finalize the page designs, with submissions normally occurring in signature-based groupings of reader spreads (sequential XPress pages in multiple-page electronic files).
3. The scanning sequence includes several subevents, including the generation of low-resolution OPI reference files and the creation of a database to manage them. The low-res files are then sent either on SyQuest disks or via ISDN directly from the sampling/scanning station to the client

site. The high-resolution versions of the images are stored on Lanman's network.

4. Once completed, page designs are sent either via ISDN or on SyQuest cartridges with accompanying black-and-white laser proofs to Lanman.
5. All files are preflighted for accuracy and completeness, and individual page components are checked.
6. Necessary trapping is determined and executed.
7. The "thin" PostScript files are imposed digitally using PressWise, which in some cases can save several thousand dollars on a single catalog.
8. The files are merged with the high-resolution files previously stored on the Lanman network.
9. Files are printed to a spooler as "thick" Postscript, with all of the digital color data embedded in the core files.
10. The thick PostScript files are either output to an Agfa Avantra 44 and proofed conventionally, or they are processed with Adobe Acrobat Distiller (the resultant files are highly color accurate due to the way Lanman controls Distiller's settings and the transforms occurring to local and remote color output devices).
11. Film-based proofs are reviewed internally and then by the client; changes are executed as required. In the case of files that have been distilled in Acrobat, they are transferred via ISDN to the client site. Some clients have 3M Rainbow digital output devices that Lanman maintains. Using scripts written by Lanman, clients' color comments are embedded in the PDF file and sent back to Lanman over the ISDN. The speed of the network, combined with the compression capabilities of Acrobat, create a unique remote approval sequence that can greatly streamline the creative/production cycle.

Thick and Thin PostScript
Thick PostScript files contain full-resolution image files. Thin PostScript means PostScript files with low-res image files and pointers (via OPI) to the high-res files. ("Thick" and "thin" PostScript are unofficial terms we've come across in several shops.)

SUMMARY

Lanman Lithotech negotiates a workflow plan up front with its clients. Because many of its clients produce hundreds of pages at a time, this strategic approach to workflow planning is critical to the smooth flow of the manufacturing process.

Artlab

Cross-Purposing for Impact

Artlab, in San Francisco, California, is one the most successful West Coast design facilities. Founder Tony DeYoung and his staff are at the forefront of work with interactive and video applications, and have garnered top clients over the years. Entertainment companies were among his first clients. Later, high-tech companies like Apple, Adobe Systems, and Kodak solicited DeYoung's expertise in developing "rich" communications vehicles on their platforms and using their software tools. Artlab, originally established by DeYoung as a consultancy, began the real-world job of applying his design skills to create professional, standalone communications vehicles.

Artlab has evolved into a production and design studio with a twist. Rather than simply specialize in video, print, or audio strategies, DeYoung mixed all three. Artlab was one of the first and still is one of the most successful cross-purposing studios in the nation. Artlab collaborated with Apple Computer on interactive demos, including those found on the AppleScript CD. Additionally, Artlab was instrumental in developing the interactive tools supplied during the rollout of the Power Mac. They were also involved in the rollout CD for Kodak's Photo CD product.

Among DeYoung's clients that participated in the PowerTools for Publishing seminar series, sponsored by Apple as part of its RISC rollout, was Ben & Jerry's Ice Cream. Its internal employee newsletter, the *Rolling Cone Magazine,* will be the focus of our workflow analysis for ArtLab.

The Project

The *Rolling Cone* project needed a two-tiered message. The first part was a "demo site" interactive version of the newsletter delivered in RGB format. The second tier was print collateral that paralleled the creative direction of the electronic newsletter and made use of exactly the same content list.

The first thing Artlab needed to do was determine what the

Cross-Purposing or Repurposing
These terms refer to a new requirement in the graphic arts field. That is, the need to use the same digital designs for different media, especially for printing and RGB media such as video or slide presentations. Questions of file size, resolution, and color space often arise and are critical. But there are no easy answers. Ideally, a single, centralized version of a graphic file could be produced and cross-purposed for any situation that arises.

deliverables would be—what exact form would be presented to the viewing or reading audience? The following is what was determined:

- A partially self-running demo (one that would require the intervention of an operator) to run during the audience-viewing portion of the demonstration. This would allow an operator to "drive" the interactive message, while a large audience was viewing the activities on-screen via an RGB projection device connected to the Macintosh running the demo software.
- A print collateral piece that would be output on a Heidelberg GTO-DI digital-direct sheetfed press. The decision to output the project to this press was made partially to demonstrate the ability to take cross-purposed digital content and bypass conventional film prepress. Secondary to that decision, but by no means a small factor, were the savings realized by eliminating the events required to output to film, manually assemble, and output the project on the floor at the Seybold San Francisco convention. Those two factors made the GTO-DI digital-direct press a logical choice for hard-copy output.

The Process

In DeYoung's mind, as is the case with many design teams, the copy is ultimately the creative—no matter what form that "copy" might take. The message—verbal, visual, or auditory—was made up of words. To that extent, Artlab assumed full responsibility for taking what Ben & Jerry's had already been saying and repackaging it for a digitally savvy audience. DeYoung's job was to market the company, so he required, and gained, client approval to control the look, feel, and content of the message.

Technology, in DeYoung's mind, allows Artlab to closely

mimic work that has been done before—without the need to completely reinvent solid creative work that already exists. Print materials, television clips, news releases, brochures, internal documentation—anything that portrayed or gave insight into the company's image of itself was collected and gathered at Artlab. In the early stages of a project like *Rolling Cone,* DeYoung tries to identify the look and feel that's best for a particular client. Simply looking through Ben & Jerry's extensive collection of graphics gave Artlab a solid starting point for the communications effort. By the same token, collecting every possible shred of existing material expands your creative options and eliminates the need to start completely from scratch.

With existing creative materials at hand, you must determine which are going to be displayed, under what viewing conditions. In this case, there were two distinct categories: RGB for screen display and slide shows and standard CMYK for printed matter. The overall project began taking shape on the RGB side of the equation, with DeYoung and his staff executing some initial screen designs, creating some linked buttons, and developing preliminary logic maps on how the various screens would be linked during operation. For example, where would clicking on a specific button take you, and how would you get to and from the beginning from any location in the program?

At this point, DeYoung was working with client-supplied content that had already been developed for print. Each component, whether a transparency or an existing printed piece, had to be scanned and then mapped to color models for computer display. The rasterizing, rescreening, and color-mapping capabilities of Photoshop in particular greatly facilitated this effort. Throughout the process, continual changes were being made to the basic design of the piece. An ongoing dialog with the client was an integral part of the evolutionary process. Work would be done, approved or modified by the client, reworked by DeYoung, and resubmitted to Ben & Jerry's. During the early approval cycles, dye-sublimation and electrostatic color output were used to provide color preproofs of the print portion of the project.

The staff at Artlab continued to collect materials while conducting extensive interviews with people at the client site. Questions focused on how the client produces its products or creates its services. The entire manufacturing cycle, including workflow analysis, is part of the information-gathering event. In DeYoung's mind, you can never know too much about the client and how and why they do things the way they do.

In the case of the *Rolling Cone* project, the Artlab designers had two primary distribution strategies to develop. One was the on-screen product, which required them to design some initial screens, buttons, and rough logic triggers. For the print side of the project they needed high-resolution imagery that fell within a gamut achievable by the press on which they intended to execute the run. Much of the material that was originally supplied to Artlab was already formatted for print. The rasterizing capabilities of Adobe Illustrator allowed them to convert existing vector artwork that had already been separated to use for video. They did lots of hand sketches of both video and print designs. In DeYoung's words, "We were creating and recreating comps at a steady pace."

During the design period, a great deal of time was spent converting traditional, painted art boards into digital data streams while trying to establish an acceptable electronic color palette that would serve both needs—calibrated RGB as well as SWOP coated standards for the Heidelberg. In essence, Artlab was attempting to simulate in a PostScript environment what Ben & Jerry's artists were achieving in paints. The Ben & Jerry packaging is highly artistic and very environmentally conscious. Unfortunately, this led the original artists to use lots of greens and reds that were outside of acceptable desktop gamuts. In the packaging industry, they simply run CMYK and two or three spot colors to achieve out-of-gamut colors. Artlab didn't have that option for this print test, which was meant to show relatively conventional four-color process being generated under filmless, direct-to-plate operating conditions.

Artlab decided to store everything at print resolutions, even though the video production would require far less space-intensive files. This choice was made to avoid having to maintain, or even generate, two different content lists until the last possible moment. The only images that were transformed to CMYK at this point were those that contained colors far out of gamut—flame reds, deep forest greens, etc.

When asked how they identified out-of-gamut colors, DeYoung said that now he might make more use of the Gamut Alarm found in Photoshop, but at the time, the function wasn't available, so he did it by eye. To DeYoung, the same kinds of colors that cause gamut problems in video or calibrated RGB cause trapping problems on press—reds, greens, bright magentas, etc. Anti-aliasing is often used on the same pieces of artwork in calibrated RGB (to provide relief from having to trap the files on press).

Calibrated RGB
Calibrated RGB is a version of device-independent RGB color space. It is useful for accurate, mathematical color conversion to other CIE-based color spaces.

Once layers became available in Illustrator, maintaining control over a specific layer's transparency solved a lot of problems for the Artlab designers. In the case of this project, when they needed to screen back a collection of images to put them behind a calendar, for instance, they simply created a big box that overlapped the necessary items, then applied a transparency filter to the entire piece of art.

DeYoung feels that learning how to solve specific problems during an actual workflow is a big part of providing services to his clients. His artists feel that every project they do makes them that much more qualified for the next one. Although many shortcuts prove very tedious to discover, they pay off in spades the next time they are needed.

The final interactive demonstration was built in Apple Script. Artlab essentially rebuilt the Ben & Jerry's information distribution and marketing workflow for an RGB delivery vehicle.

One of the major challenges of the project came when

Artlab attempted to use the same exact objects on-screen that had been printed on the Heidelberg press. Working with direct-to-plate color was new to Artlab, and it took a lot of manual adjustment to individual images to bring the gamuts within an acceptable range, while maximizing the output from each respective device.

In particular, Artlab found the sign-off process difficult. They essentially went to plate without any contract proof. Artlab felt that, in large part, the project came off successfully at the end because everyone was so cooperative. The artists pulled early press proofs and also output the file to a color digital printer. They immediately encountered color-shift problems on the digital printer. In particular, one of the key background blends printed with a major color shift.

The background in this image is a simulated "sky" containing a lot of noise. At about the 10% range of the blend, the plate simply lost the remainder of the gradation. Instead of continuing the diminishing tonal ramp, the rendering produced a solid line. The solution was to grossly roughen the range using a Photoshop filter and thus get somewhat better results on the final press run.

Immediately after the first run of 1,000, the client came back and requested 10,000 more. By this time, Heidelberg had upgraded the plating technology to a newly released product from Presstek, and the second run, in DeYoung's words, "ran perfectly. Astounding. The ramp problem disappeared."

In discussing the project after the fact, the people at Artlab felt that they had spent far too much time, as designers, dealing with color shifts, models, transforms, and other noncreative issues. On the other hand, their creative intent required them to explore and push the edges of the production model to achieve the desired results (kind of a chicken-and-egg thing). Therefore, they don't think it was a bad thing; not to mention it's what sets them apart from the other guys.

The current state of the market seems to indicate that having a high degree of technical knowledge not only makes you a more effective user of your tools, but also gives you a competitive advantage when working with highly technical clients, such as Apple, Kodak, or Adobe Systems. In the end, it's a trade-off. Creators would much rather spend all their time designing, but they don't have that option. The tools they use haven't evolved quite far enough to minimize the production-specific issues they have to deal with on projects like *Rolling Cone.*

Artlab and AppleScript

Since Apple is one of Artlab's major corporate clients, DeYoung was exposed to AppleScript very early in its development. He routinely uses it and related scripting utilities in an attempt to automate any processes his team can identify as being "automatable" and applicable to other projects.

AppleScript comes into play on many of Artlab's projects. As part of its Power Mac rollout program, Apple hired Artlab to create a compact disc containing all of the speakers' slides, some of their scripts, video from the show, product data sheets, and other digital "goodies" from the seminar series (PowerTools for Publishing). Since most of the slides were originally done in Persuasion and needed to be rescaled, anti-aliased, and organized into a Macromedia Director document, one of the first event "assignments" was to build a script that would:

- Launch Persuasion
- Write the document out to a series of TIFF files
- Launch Photoshop
- Open the documents one at a time
- Resize
- Anti-alias
- Save

- Launch Director
- Open the folder
- Import

All of these events were triggered within an AppleScript routine, which used the folder contents as a data source, with each "record" (each document to be processed) being represented by a single field: the file name.

Since the client wanted a set of slides as part of the final product inventory, the script generated high-resolution slides that were output to a high-resolution slide recorder. In DeYoung's words, the workflow "worked flawlessly; because of the slide requirements, the files were huge, and perfect."

DeYoung feels that scripting events will continue to be a large factor in his studio's future success. (It is interesting to note that since this case study took place, Adobe ScreenReady®, which performs the first seven steps listed above, has been launched.) One of the firm's primary research and development goals is the movement of its design and technology skills onto the World Wide Web.

Summary

Design firms have serious workflow issues, especially as they get more and more involved in the production effort. As with Artlab, this is especially important as designers repurpose content for varied media. Cutting-edge creative firms have to take the lead in pushing the tools they work with to their limits. This often means that they know more about what the tools can do than the producers they send files to. And, like the producer site, a creative environment must also develop a workflow approach that may include some production tasks. As the artists point out, time lost figuring out correct methodologies usually pays back on the next job because they are that much smarter going in.

Buchanan Printing
Strategic Preflighting

Buchanan Printing in Dallas, Texas, is one of those rare printing companies that operates its prepress services as efficiently and profitably as its printing facilities. The company excels especially in preflighting and estimating jobs, for which it has developed winning strategies—strategies that pay off in terms of both throughput and customer relations.

The company has been operating as a printing house since 1957. The current owners, Lyn and David Johnson, bought the company in 1980; under their leadership, Buchanan has grown from sales of $350,000 with five employees to $11 million in sales and 105 employees. The printing is strictly sheetfed, with five presses, including a new 40-inch Heidelberg 640 six-color press. With a background in productivity and retail engineering, Lyn, who manages the production facilities, easily made the transfer into electronic page composition.

Their move followed the usual paths, from traditional typesetting, to acceptance of customer disks, to the purchase of an Agfa 9600 as a transitional machine, and finally to a full-color production system. Now the company has eleven Macs, Agfa SelectSet 5000 and 7000 imagesetters, a Crosfield 626 scanner with a MagnaLink feed to the desktop, and a LeafNet system for critical color movement. For proofing there is Matchprint, an Iris 4012, and a CalComp plotter for inexpensive color.

Overview

Buchanan started a color prepress department from scratch. But Lyn Johnson notes that it wasn't easy. "Most people think the learning curve is six months, but it's more like 18 months." At this point, they supply the vast majority of their own films. Outside-supplied films made up 30% of the print-house volume two years ago—now it's only 10%. They normally don't do prepress work for other printers; after all, "we don't want to make another printer look good."

At Buchanan, prepress charges are itemized separately from printing, even though most jobs involve both services. The dividing line in the shop comes with the production of composite film. Johnson says that this approach makes it easier to estimate jobs realistically, rather than trying to fudge figures. "I also want to show people I'm competitive on both sides, so it becomes an effective tool for marketing [our services] to customers."

Lyn Johnson feels that her company, by offering prepress services, has a big advantage over service bureaus and trade shops. "We try to eliminate our competition by trying to ensure the quality of the finished piece and remove the pain of babysitting the design through several vendors. We know exactly what line-screen value to use on a photo, the precise trapping requirements for our processes, the appropriate bleeds, the correct stock selection necessary to improve definition or to create the look the designer needs. We also provide telephone support while they design the piece, if necessary."

Buchanan's concept of "architecting" a job (workflow design) in the prepress department involves understanding how the printed piece is going to come together on press. The Johnsons judge productivity not by the pieces of film imaged at the end of the night, but rather by whether the film produced will make plates that run the first time.

This architecting process has an advantage in the printing department. "We have taken press OK from an average of an hour to a half hour," says David Johnson. "We've also cut plate problems in half. [Some plates have natural flaws, so there will always be some bad plates.] Usually we run four-color process and a fifth layer as type, since that's what people change the most at the last minute."

Buchanan comes by their "architecting" competence naturally, by the judicious use of the talents of their best (former) strippers. (They once had seventeen, and managed to keep the best eleven, the skilled ones with the right attitude to assume a

Retraining Srippers
Buchanan invests heavily in training. Resources are set aside for conferences, classes, and, most important, time. They treat training as a line-item expense and deliver it as part of the cost of maintaining each employee.

variety of jobs.) According to Lyn Johnson, "In the old days we would have gotten composite film from others. Our journeymen would look at the film, see potential problems, but very often would not fix it until the customer saw the first proof. At one time, the mistakes would have been seen as 'not my problem,' but now our workers have to be more analytical and more diagnostic about how the job is laid out. They have to check trapping, imposition, gutter jumps, and bleed. They stop thinking of themselves as film assemblers; they have been elevated to film diagnosticians."

Buchanan tries to create a positive environment for catching problems before they get to the film stage. Anyone who catches a real problem, one that would cause a job to be rejected during proofing, on press, or by a customer, can stop a job. "Everyone has the right to pull a job," says Johnson. "I didn't want anyone to feel that a job couldn't be pulled." Points are given for catching a problem; with so many points the employee earns a day off. The number of points required differs per job assignment, since, for example, the proofing station has more opportunity to catch problems than does the scanning operator.

Buchanan wanted to develop a constructive way to analyze and fix problems. The person who makes the mistake has to sign off on it when it is caught. But assigning blame is minimized. Finding problems has become a friendly competition. As it is, virtually every problem is tracked down and signed off.

But in order to get over the psychological difficulty of admitting guilt, Buchanan now has a category called "Other Operator." This is an oblique way of signing off on an error and admitting guilt, but deflecting the personal onus a bit. It is clear from the documentation who made the mistake, but this makes it a little easier to learn from mistakes, since it defuses the self-consciousness a little. The joke around the plant, according to Johnson, is that "if we ever catch this Other Operator person who's screwing up all these jobs, we're going to fire him!"

Film Wastage

Another way that Buchanan keeps tight control on production is by monitoring film wastage. The former camera man has turned into a quality-control operator with responsibility for spoilage and wastage. Wastage means unused film. Spoilage means film that has to be discarded because of imaging problems. Spoiled film is nonchargeable (AAs, for example, would not constitute spoilage).

The people at Buchanan were shocked to find that once they started checking, they had amazingly high rates of wastage and spoilage combined, bordering on 50%. One main cause was easy to correct. Every job on their Agfa SelectSet has a 9-inch head and a 7-inch tail of unimaged film, with a 2-inch gap between pages. A single 8½-x-11-inch page ended up taking 23 inches of running film. They had to get the factory settings reduced to 8 and 6 inches for head and tail, with a quarter inch between pages. This change required Agfa to adjust the settings mechanically. They also changed operating procedures. Now they run the imagesetter only when they have 25 pages or more to rasterize (or every two hours, whichever comes first), cutting the head and tail loss to the beginning and end of the whole 25-page sequence. This means they are effecting enormous wastage reductions (jobs almost never wait, since the imagesetters are almost maxed out, running at 70% utilization, twenty-four hours a day).

They have also reduced spoilage by preflighting and by catching errors before they get to the imagesetter. Combined with their wastage savings, they now have about a 9% film loss on a weekly basis, and their best week ever had a 5% loss. This is an amazingly low figure. (Agfa documents film loss near 18% as standard.) The savings come not only from the film saved over the year, but also in the time and trouble saved in catching errors early.

One related point—the SelectSet 5000 and 7000 are calibrated to each other so they are in synchrony at all times. If one page of a separation needs to be rerun, it can be run on either machine with exactly the same results. (They calibrate the machines several times a day.)

Form 5000/7000

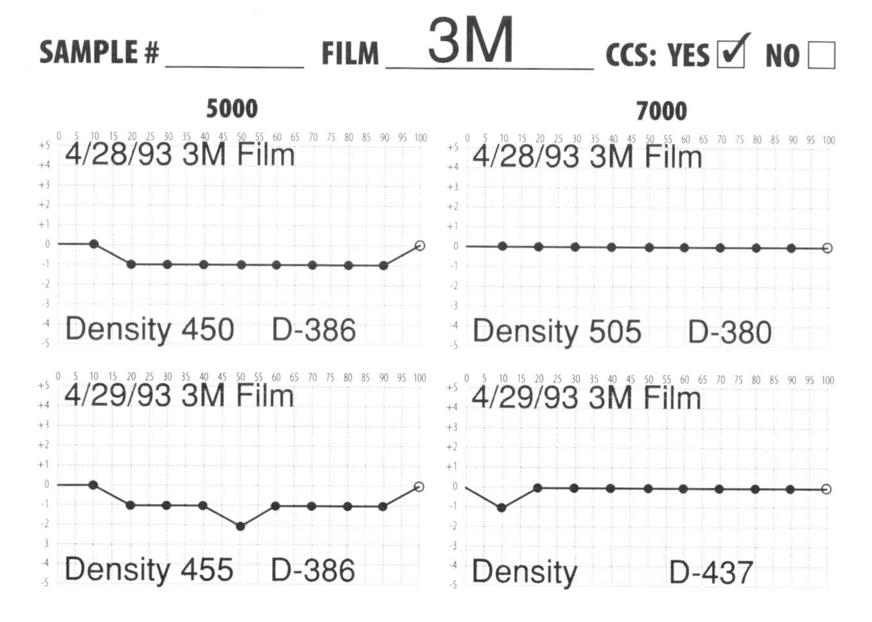

Preflighting

Buchanan is unique in its extensive preflight effort. It sees preflight as input quality control, which is more effective than output quality control. The Johnsons add a $100 diagnostic fee in the estimate, not as a line item, but as part of the overall job cost. Their attitude is that if they lose a job for $100, shame on them. In the preflight, a trained preflight operator runs up a detailed estimate of the job that is compared with the initial estimate from the salesman or estimating department.

To develop good preflight operators, the company looks for a Mac-competent designer or production worker who knows the principal applications. Then they train him or her for three months in stripping, three in estimating, and two months are spent in production planning. So their preflight operators have a solid eight-month apprenticeship before starting, along with graphic arts experience.

There is a four-hour-turnaround guarantee on all preflight. Once a job is physically submitted (usually on a SyQuest), the preflight engineer starts to analyze the job, noting problems and missing items, and assigns charges for producing every element, from scans to trapping, including conversions, files, fonts, bleeds, and degradés. The preflight engineer opens the application file, runs PostScript comps in color, and brings a wide range of expertise to bear on whether or not the file will run. He or she also determines when fixes are needed. Within a few hours, a preflight-based estimate is ready.

This new estimate is reconciled with the document produced by the salesperson or estimator. Any amendments are made, though the preflight department's comments are usually accepted. The report is then used as a vehicle for communication. Instead of being charged additional fees at the end of the job, the client can fix problems themselves, drop problem elements, or pay Buchanan to fix them. All this minimizes surprises at the end, keeping clients happy while not slowing down the overall workflow.

Preflighting to-do form

JOB # 42434

	DONE	OPERATOR
7 4/c transparencies for scanning. 5 of them need clipping paths created. All are 3" x 5" or smaller. Need to place in Quark	○ Yes / ◉ No Time 8 Hr(s) Scan cost 1 $600.00	
Need to verify what is and what isn't varnish plates. If we need to create varnish for 4/c, that will be additional time	○ Yes / ◉ No Time 3 Hr(s) Scan cost 1 $225.00	
Need to several elements in Quark and Freehand. Does small text in	○ Yes	

In addition, the preflight department generates hard copy of text at preflight. They fax it to independent proofreaders, who are instructed to look for critical errors (spelling mistakes, missing sentences, and so on). Minor areas like commas and word choice are ignored. The customer can fix typos, or Buchanan will do it for a fee. This, notes Lyn Johnson, is not a profit center. It does, however, provide an extra service at a break-even cost. After all, the prepress house is blamed for embarrassing errors even when it's not its fault.

Preflighting has a great track record at Buchanan. It reduces mistakes at the imagesetter, speeds throughput, keeps customers informed early, and allows for no-apologies billing. It belies the myth that "there's no time for preflighting" by making for a smoother workflow and a low percentage (20% or less) of rework.

Summary

Buchanan Printing anticipates problems on press from the moment the page layout file arrives. It emphasizes preflighting in its workflow approach and tries to catch errors as far upstream as possible. They feel that preflighting is not something you can buy in a box; rather, it is a conscious set of strategies used to deal with customers and their work. Buchanan especially focuses on bleeds, traps, and page crossovers.

Since we met with the Johnsons they continue to modify their workflow. For instance, as CSRs become more familiar with preflight processes, they are given more responsibility. Buchanan has shown that preflighting is the key to profitablity and that it leads to increased customer satisfaction.

CASE STUDY

ST. PETERSBURG TIMES
Working Cooperatively

No other creator/producer environment matches the pressure of a large daily newspaper. In this case, the *St. Petersburg Times* (St. Petersburg, Florida) maintains a level of editorial and graphic excellence that consistently places it on the list of the nation's top ten daily newspapers.

For years, the *Times* has been known for the innovative use of high-level graphics and four-color process in the production of the paper. For the purposes of our interview and analysis, we selected one of the more design-intensive departments found in the publication: the Floridian, a Sunday special-interest section that typically features stories of interest to local readers.

Artist Victor J. DeRoberts came aboard as features designer in April 1992. His previous experience was largely in news graphics, having held positions with the *Naples Daily News* (Naples, Florida) and thc *Memphis Commercial Appeal.* At the *Times,* he's responsible for, among other things, the *Floridian,* the *Weekend* magazine, special food sections, and other "socially related" features. His feature designs are well known in the industry.

Since DeRoberts's designs are routinely design intensive and always make use of high-resolution color images, we could track several important aspects of the design and production process. In the newspaper industry, unlike many typical creator/producer relationships, the entire event sequence takes place under a single, focused management. Both creator and producer work under the same roof. Here we also have a high-level sample of "desktop" color being used in a time-intensive, colorful, design-driven workflow.

The Process

One of the first questions we posed to DeRoberts was how a piece comes together: who decides the content of each week's articles? Who knew about the "Ghost Orchid," and how did the story make it into the paper?

The "Ghost Orchid" story started with a desire by senior staff writer Jeff Klinkenberg to cover this nature story. The story is not only about the orchid, *pollyrhiza lindenii,* a "leafless plant that isn't all that attractive except for the three weeks or so that it happens to be in bloom." The underlying story concerns a highly endangered, unique area of the greater Everglades system called the Fakahatchee Strand. Almost completely decimated by cypress clear-cutting in the '20s and '30s, the area is host to hundreds of plant species found nowhere else on the planet.

Features manager Chris Lavin told us that with more senior and trusted writers like Klinkenberg, the decision to put the story in the paper is almost perfunctory. The writer knows what the reader will like and is fully capable of managing the project from an editorial standpoint.

Lavin has a budgeted amount of space for his features sections, and begins to determine the editorial priorities and ad balances roughly a month before the run time. He must remain flexible, of course, to accommodate late-breaking, very important issues, but generally likes to remain ahead of the press run by about four weeks. About a week after he's begun to collaborate with the editor about the content and has assigned the visual requirements to the paper's photography group, the story shows up on the budget listing, which estimates the time and expenses required for DeRoberts's design assignments.

In most cases the photography department manages the assignments coming in from the many departments and sections within the paper. In this case, Scott Keeler, a photographer that works in the paper's Clearwater branch, had expressed an interest to work with Klinkenberg. Lavin knew this and told Keeler and Klinkenberg to head into the swamps.

The editorial comes together simultaneously with the visuals. Keeler might shoot as many as 200 images, and upon development and visual analysis, reduce the "bucket" of images to eight or ten. He does his own scans. Although there is a department within prepress dedicated to executing the hundreds of scans required per day, and for managing the paper's Leafdesk image database, certain skilled photographers are beginning to execute their own work. Since the paper (currently) runs at a 100-line screen, the resolution requirements for the visuals isn't so high as to be outside the capabilities of the paper's extensive Macintosh and Windows systems. As a subprocess in the scanning sequence, Keeler crops the images as well.

Increasingly, the process of editing photography and images is done on high-speed Macintosh systems. In the time-sensitive world of the daily newspaper, this doesn't always allow for segregation of skill sets. Photographers like Keeler are dealing with such production-specific details as undercolor removal, gray component replacement, and other fairly high-level decisions that really have nothing to do with the visual itself, only with how well the photographer can expect the images to reproduce on press. (In the opinion of the authors, making reproduction decisions should be removed or diminished from the creative event sequence. Instead, it should be a production event—not a function of the design process.)

The copy and visuals come together as a cooperative effort with DeRoberts, Keeler, and Klinkenberg reaching a general agreement about final copy (how much or how little), and which images DeRoberts feels will best suit the layout. There's a balance between the editorial and the design-driven nature of such features and ultimately, the team has final say. With the exception of proofreading, and a quick look by Lavin, this piece ran as the team approved it.

While DeRoberts is putting together the Quark XPress mechanical, the images are stored on one of the paper's dozens of servers. The system's responsiveness is never an issue. "There are about 15 or 20MB of image files on the page," says DeRoberts. "There's no problem at all placing or working with images stored on the servers. I hardly notice any delays. Traffic isn't an issue since we have so much horsepower on the system."

The editorial is originally entered on a proprietary editorial system in use paperwide. The files are moved to the Macintosh during the design process, and final edits are made directly in the layout. This reduces the amount of time that the copy has to move back and forth from writer to page and back to writer. As is often the case, the team is very used to working together, and workflows are relatively stable. As in any high-volume situation, there are always problems, but there aren't any endemic problems without apparent solutions. The paper works around what can't be done perfectly.

Once the pages have been completed, they're normally output to one of several different proofing devices. At this point, they're still in PostScript. Before sending the job into one of the predetermined output pipelines, the team normally saves a lower-resolution version of the image in EPS format, preserving the design and look for subsequent reference or research. They've begun to make use of Acrobat to accomplish this task.

Once completed, the page is "printed" to spoolers maintained by a specialized group within Information Management: the Prepress Group. Department manager Tom Heyer is, at this point, maintaining a spooler/server operating on a Sun SPARC10. The spooler manages multiple output devices, including proofing printers and four high-level vBit RIPs (vBit is a proprietary bitmap marking language owned by Hyphen, a large player in the newspaper reproduction industry). The RIPs can feed any of the site's three III 3810 devices (one for RC paper and two for film) or the printing plant's remote III 3850 broadsheet imposition setters.

PDF
Portable Document Format creates a distilled or simplified version of a PostScript page that includes embedded fonts and illustration(s). The file format prints and displays on remote printers and monitors exactly as it does on local systems. The compression capabilities of PDF make it highly suitable for passing files on the Internet.

Further, recent development efforts have resulted in a home page on the World Wide Web. There are extensive plans for increasing the paper's presence on the Web. Clearly the publishers recognize the growing need to cross-purpose their content in forms other than print.

Another area of interest in the (many) workflows of the paper is its recent decision concerning Adobe Acrobat. Due to the wide variables found in incoming ad files being prepared electronically by outside agency sources, the paper intends to standardize on PDF files and will base submission requirements on PDF starting in 1996. Currently, a lack of OPI referencing and accurate color handling in Acrobat remain an obstacle, but the many problems they encounter in receiving outsourced electronic files demand a long-term solution, which they feel Acrobat will ultimately provide (see Chapter 8, "What's Next").

Summary

The workflow strategy at the *St. Petersburg Times* hands more production responsibility to the image editor, rather than to a pure production operator. At the *Times,* linking production tasks to aesthetic ones saves time and minimizes errors, thus enabling maximum quality in the shortest turnaround. The tight deadlines and heavy demands of a newspaper schedule require that narrow departmental roles and distinctions be telescoped together. Establishing a usable workflow that allows form or production issues to be handled by creative staff is critical.

Graphics Express
Quick-Turnaround Artists

Graphics Express, in Boston, Massachusetts, is no ordinary service bureau. In six years it has grown from three partners with a LaserWriter and a Macintosh to a $10 million-per-year business with 11 imagesetters, a sophisticated drum scanner, color output capabilities, and a staff of 150. Graphics Express is a serious competitor to some of the biggest trade shops in the area. The company is still growing at an amazing rate (5% per month). It is building a multimedia department, expanding quarters in downtown Boston, and the company is one of the first sites for the Chromapress, Agfa's version of the Xeicon digital color printing press.

Despite its sophisticated equipment and high income, Graphics Express is still a service bureau at heart. Rick Dyer, one of the principals of the company, is proud to point out that elaborate preflighting is not a realistic option for them. The vast majority of files (over 200 jobs a day) come into the shop in the course of an afternoon and are sent out the next morning. The receiving department takes the disks, makes sure that the files are there by checking them quickly as they are copied to the server, and then enters data on the Job Info screen, which creates an electronic job ticket for the job. Any unreadable or empty disks can be found at this time. This is the extent of preflighting that occurs.

Operators first check the file for printability, then print it, usually first to the laser printer, then directly to an imagesetter. Any problems that crop up are handled by the job's operator. Serious difficulties may be handled by the shift manager. All of the operators are trained internally in applications and systems, and only a small number of jobs require extra effort by a more knowledgeable shift manager.

Because Graphics Express provides overnight turnaround, preflighting is not an option. The hundreds of jobs that pass through the shop each evening cannot be processed by a separate preflighting department. Besides, more often than not, clients have gone home and are unavailable by the start of the evening shift, when most jobs are processed. The penalty for missing elements or fonts is that delivery time gets postponed, and the day shift handles the cleanup task.

Problem Tracking

With so many jobs coming through so fast, the most critical issue at Graphic Express is not preflighting, but job and problem tracking. To make sure that the current job status is understood and communicated as shift changes occur, and that phone calls are returned, Graphics Express uses a 4th-Dimension-based database that tracks all jobs. This allows jobs to be searched by client or by status, and even produces shipping information. Management can generate reports of pending jobs at the beginning of each shift. On completion, job information is fed to accounting, where the data is used for invoicing.

The Job Info screen is filled out when a job comes in. Each job has a receiver (who opens the jobs, copies it to the server, and enters it), a production supervisor, and a quality-control person (who checks to see that the job elements are there and that the job is correctly reproduced). A scrolling notes field allows for problem documentation. In this example, the bane of all follow-up attempts, voice mail to an after-hours contact, is at least documented.

The Job Problem screen allows quality control to document each printing problem and the technique used to solve it. The most common causes of problems are categorized and there is ample room to document them for further clarification.

The Job Rerun screen allows for complete documentation of jobs that have to be rerun. Some reruns are initiated internally, while others are requested by customers. The often delicate decision of who pays for the rerun is handled by managers. As this example illustrates, there can be many misunderstandings and miscommunications that lead to several iterations of reruns. Full documentation of these problems allows for defensive billing and clears the air when the finger-pointing starts. For this reason, documenting the exact times when jobs were attempted to be processed, or when calls were made to the client, is extremely valuable.

In the context of a three-shift, seven-day operation, with scores of operators and service personnel, this elaborate job documentation system has been critical to the success and rapid growth of Graphics Express.

Summary

When a company like Graphics Express is bombarded by jobs demanding quick turnaround, workflow demands are critical. It's especially important to find methods for dealing with pending work, since formal preflighting is impossible. Instead, an efficient and quick job entry procedure and a tracking system database is the workflow strategy that Graphics Express uses to handle its enormous workload. The hand-off from the evening shift, which processes most of the work, to the morning shift, which, using well-documented error reports, takes care of mop-up, is the crucial event of every busy day.

CASE STUDY

Gamma One

Team Strategies

Gamma One, located in North Haven, Connecticut, is a major color separator specializing in catalogs. It has been struggling with workflow in the context of Total Quality Management (TQM) for a long time. Bryon Ramseyer, principal and head of the production effort, is a fervent believer in empowering employees and following TQM philosophies. In general, the company has done quite well but in recent years, with tighter competition, tighter profit margins, and tighter schedules, Gamma One is under more of a squeeze. Ramseyer will be the first to tell you that earlier TQM efforts have had only mixed results.

Over the past few years, however, Gamma One has managed to improve its situation dramatically. First, it brought in Vince DiPaola as vice president—one of the top technical minds in prepress today. Vince has had major success in integrating the company's network and server strategy, training clients and configuring client sites, and building the high-speed telephone links and servers to keep all this traffic humming smoothly. Because of his work, the plant has taken on more work without having to add new equipment or personnel.

Second, Ramseyer and his staff have instituted a serious new workflow strategy that combines the best parts of TQM and yet is flexible enough to allow for sustained improvements in productivity and quality in the rapidly changing environment of prepress. Combined with a new data tracking strategy and a refocused sales effort, Gamma One continues to grow, profit, and produce high-quality work, while maintaining a relatively relaxed and humane work environment.

Teams

The most significant workflow change at Gamma One is team orientation. Until this change was implemented, Gamma One had a general production staff and assigned incoming work on a first-come, first-serve basis. CSRs worked in a totally separate department in the administrative part of the building, several hundred steps, and several walls, away from the work floor. Job scheduling and shift assignment was done by a cadre of middle managers, and the frequent job delivery crises were handled in meetings with participation from middle and upper management.

Now there are seven production teams of about 13 employees each. Each team has a CSR, at least one experienced color expert, and a number of people competent at page assembly and proofing. Each team is assigned to one or more clients. They determine their own schedules and resolve most production problems on their own.

With the exception of scanning (which is a specialized group whose services are "hired" by the other teams), the team performs all required services for the customer. These include: color correction; creating silhouettes, making shadows and tints; trapping; final page assembly; proofing; final output; and shipping. Some resources (workstations, for instance), are "owned" by the groups and others, such as imagesetters and proofing machines, are shared.

Each team "buys" color from the scanning department. Together, they agree on how they want the scans delivered, how much color correction is needed, and which are to be made on the scanner versus the workstation. The scanning department has two Linotype-Hell 3800 scanners, with prescanning stations attached—a productive machine of superb quality. As company principal Bryon Ramseyer says, "If we could get similar quality scanners for about $10,000 each, each team would have a scanner. Right now, we see no way to get the quality we need except from a high-end drum machine."

Each team has a leader whose main responsibility is to see that certain tasks get done. This may involve prioritizing and tracking work, and checking that long-term objectives are being met, as dictated by the needs of the team. The leaders are chosen by the group, with some guidance from management.

Leaders come from all disciplines—some are CSRs, some are Mac experts, and some are color experts, depending on the dynamics and personalities of the team members. The group leader is not a manager, nor does the position earn a higher salary. "You could walk into a team and couldn't tell who is team leader," notes Frank Clare, one of the team coordinators.

Responsibility

According to Ramseyer, "The team is responsible for customer satisfaction. Their main objective is to accept responsibility for quality and on-time delivery." It must be stressed again that the teams are basically self-directed.

Teams work with one or several clients. They tend to develop peer-to-peer relationships with their production and design counterparts at the client site. In some cases, they are like external employees of the client company. For example, one team works with Stanley Tools. Each team member gets a Stanley T-shirt, they all visit the plant—they even keep some of the tools supplied for photo sessions. Most important, the teams understand the customer's needs and style. Because of this unique customer closeness, teams can diagnose problems before they happen and arrange to fix the problems, either directly or through the sales and training staffs.

Salespeople usually interact with more than one team; they have to develop a close rapport with all team members assigned to their clients. It's in the salesperson's best interest to keep the team, or teams, that serve his or her customers happy and informed. This minimizes resentment that sometimes occurs between production personnel and salespeople. At Gamma One, it's clear that everyone works together.

Teams schedule their own work. This includes weekend and night shifts when necessary. Their main (indeed only) concern is to deliver high-quality jobs on time with reasonably high productivity ratios. If one team is crunched, they can "borrow" time from other groups that may not be as busy. On the other hand, if one team's workload is light, they can offer team members to busier groups. The teams work with each other to schedule the use of shared equipment and negotiate with management for their own equipment needs (upgrades to their Macs, for example).

Each team determines how the work area is laid out, and defines its own goals for improvement. Teams can, within reason, restructure their environment. Jobs are micro-managed from within the group—which files gets processed when, who takes care of proofing pages, whether we accept a scan or send it back, who silhouettes an image. All micro-functions are tracked on paper. A team can ask for assistance from management when serious problems arise, but otherwise management is not normally involved with day-to-day procedures.

At Gamma One, cross-training is emphasized within groups. For instance, CSRs are trained to do more than administrate. They often have other production skills and can pitch in and help get a job completed. As Ramseyer points out, there is an "overhearing factor" that is critical to productivity. Because they sit near the people doing the work, a CSR hears more about a job's status and specific problems than any number of meetings could tell them. (One side note: it's sometimes hard to get things right the first time. In certain team areas, the cubicle walls were too low, and there was too much noise. So they got higher dividers. Now the work area is less open and less overhearing takes place, but it is saner.)

At the end of a job, each team goes through a postmortem. Why did this image require a dozen proofs? Why was this proof rejected? How can we improve the process in the future? These postmortems tend to focus group directives and help define continuous improvement strategies.

Team Coordinators

Gamma One has three team coordinators, each with a different specialty. Mike Morin is a Mac specialist with a sales background. Ronnie Buczynksi is the company controller. Frank Clare is a color expert. These three don't "manage" the teams; rather, they act as facilitators and communicators, to make sure that the teams have the resources they need.

Early each morning the three do "doctors' rounds." They visit each of the groups one by one. The teams that need resources make claim to them. By 9:00 A.M., the resources are distributed. When groups ask for additional personnel, it's usually in color correction or proofing. Typically, resource conflicts peak from about two or three in the afternoon until FedEx time. As in most trade shops, it's usually over the use of imagesetters; because of differences in screening technologies, they can't switch between Lino and Agfa output in the same job.

The teams are highly flexible, according to Clare, thanks to their depth of experience. "Recently we had to do a job with conventional stripping. It was no problem; we still had expert strippers on the team. We hauled the tables and frames out of storage. We still have expertise for that in each group."

Team coordinators are responsible when a group is seriously in trouble, which has happened on occasion. The coordinators helped the groups restructure themselves, rather than laying down the law. As Clare told us, "Two groups had problems; we decided the teams need to be restructured. We also had to get the right mix on each team: one was too heavy on Mac operators, the other too heavy on DaVinci. We talked with all of the team members, and luckily they all agreed to rebuilding the two teams with better balance." In picking new leaders, again, the coordinators act as agents, not bosses. "We get groups to identify what the team leader has to do, then see if anyone wants to take it on."

The coordinators themselves were trained before they started coordinating. As Clare explains, "I was part of the CMT

[Change Management Team, the group that helped prepare the plant for the team approach]. Agfa came in a year ago and gave us some training based on what they do in their Wilmington, Massachusetts, manufacturing group. In all, we got sixteen hours of guidance, four hours per day." Subsequently, they have received help from Agfa in training the whole plant.

The mornings' rounds take a little over an hour. After that, the coordinators still have things to do. They talk with senior management about common resource problems and assign resources, if necessary. They also facilitate team meetings: each team has a general meeting each week, for one to one and a half hours, to work on process improvement objectives. The coordinators attend and help run the meeting. They help facilitate any general group requests. In addition, the coordinators, along with team leaders, meet regularly (the so-called Navigator group) to chart overall corporate process improvement.

The coordinators also spend time doing work within their own specialties. For example, Clare still works on some of the knottiest color files. He is leading the current effort to rethink ways to better communicate color expectations to the client.

Think of the coordinators as enablers—not bosses or middle managers. Again, their role is to help coordinate efforts among the teams and to help them better realize their own objectives.

Summary

Sometimes rethinking the entire model of the corporation can end up being an effective workflow strategy. Gamma One combines continuous improvement with a real sense of responsibility to produce quality work and on-time delivery at every level. This is accomplished using a team approach, which hands power and responsibility to every member of the team. In addition, Gamma One maintains an ongoing technical effort to upgrade equipment and tracking tools. They have changed from a company wary about taking on new business—since it made their existing workflows even more complex and less profitable—to a company that is confident to take on the world.

Case Study

West Marine

Power Tools for Workflow

While we don't think that workflows come out of box, there are a few software tools that, in the proper circumstances, can make a large difference in productivity. The package in this case is CatTrax, a product of DCS, Inc. This product greatly improves catalog workflow production, both for the publisher and the prepress shop.

The publications department at West Marine, Inc., in Watsonville, California, was in a state of perpetual panic. Not only did the nautical products firm have to produce a huge (500+ pages) color, retail-oriented master catalog for its wide array of boating supplies, but it also had to produce several smaller direct-mail catalogs. In addition, the publishing department was responsible for a number of related projects, including ad inserts and mailers. In total, it produced over 1,400 pages a year. The nine-member department (including administrators, writers, and photographers) logged many stress-filled hours.

As in many art departments, all of the design was created on the desktop. The pages were laid out in Adobe PageMaker on in-house Macs, and low-resolution scans were produced on an in-house scanner. The pages (on disk) and original images (transparencies) were then sent to trade shops. They scanned, separated, and cropped the images and finalized the pages. Proofs and corrections were made, and then the pages were finally imaged. Many pages were produced in black-and-white to save time and money.

Laying out all those pages was not the only problem, according to publications manager Randy Elwood. Coordinating the work with a number of outside vendors was a monumental task. Last year, West Marine split the master catalog between four color trade shops. There was just too much work within too narrow a window to give to one shop. Aside from the normal problems of producing the catalog, the design team had to coordinate the distribution of files and color, manage proofs, and track corrections from all these subcontractors.

Software Savvy

A few years ago, West Marine signed with a single prepress vendor, Digital Catalog Solutions (DCS) of Westlake Village, California. The company set them up with a proprietary solution called CatTrax, which they invented to ease catalog production, both at the publisher and at the trade shop. (The CatTrax software was created by DCS, Inc., of Atlanta, Georgia.) Using CatTrax, West Marine saved enough time, money, and personnel to publish a 700-page master catalog, with four to twenty color photos per page, and is knocking off mailers and smaller catalogs with ease. The yearly page count is steadily going up, as management finds new uses for CatTrax. The increased production was accomplished with no increase in staff nor in prepress costs.

The DCS Workflow

DCS's Principal, Barry Rickert, has been using PostScript color production since the pioneer days. Gradually, he developed an advanced solution for catalog assembly that allows publishers to make use of standard Mac software including: Quark XPress, Adobe PageMaker, Adobe Photoshop, Macromedia FreeHand, and Adobe Illustrator. DCS scans the images at high-res (on a Hell scanner) and, when needed, creates silhouettes and drop shadows. Low-res versions are available for the catalog's layout group via high-speed phone lines. The layout group can crop, size, rotate, and place images. Using OPI substitution, pages are proofed locally, and then, when the final pages are ready, they are output and proofed at the DCS facility. DCS takes care of trapping, imposed printer's spreads, and other production-oriented issues. The transfer of files between client and DCS is done over phone lines. (The proofs and the photo originals, of course, must be sent by overnight delivery.)

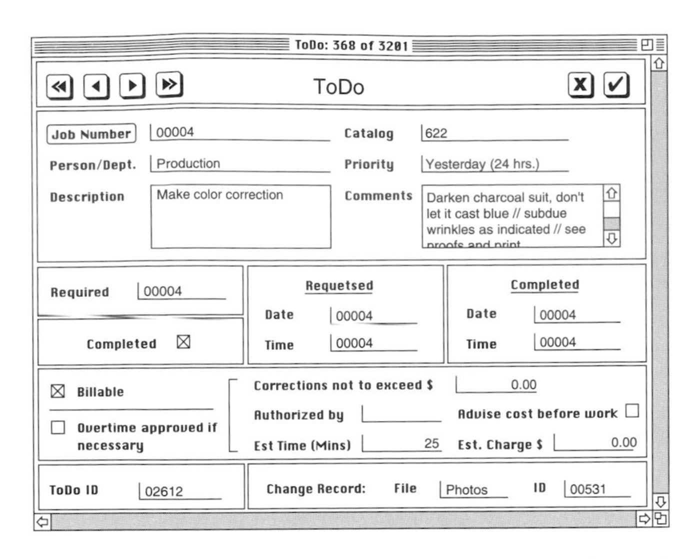

ToDo: 368 of 3201

ToDo

Job Number	00004	Catalog	622
Person/Dept.	Production	Priority	Yesterday (24 hrs.)
Description	Make color correction	Comments	Darken charcoal suit, don't let it cast blue // subdue wrinkles as indicated // see proofs and print

Required 00004

Completed ⊠

Requetsed — Date 00004 — Time 00004

Completed — Date 00004 — Time 00004

⊠ Billable

☐ Overtime approved if necessary

Corrections not to exceed $ 0.00

Authorized by

Advise cost before work ☐

Est Time (Mins) 25 — Est. Charge $ 0.00

ToDo ID 02612

Change Record: File Photos ID 00531

Working this way, the layout group sees how the final pages will look much earlier in the cycle, so they can concentrate on design, rather than production issues. Image pickup and reuse is a snap, since all the images are arranged by product code and are always on-line and coordinated with the high-res database. Elaborate tools for tracking the current status of every image and every page alleviate the administrative nightmare of managing large jobs with thousands of images.

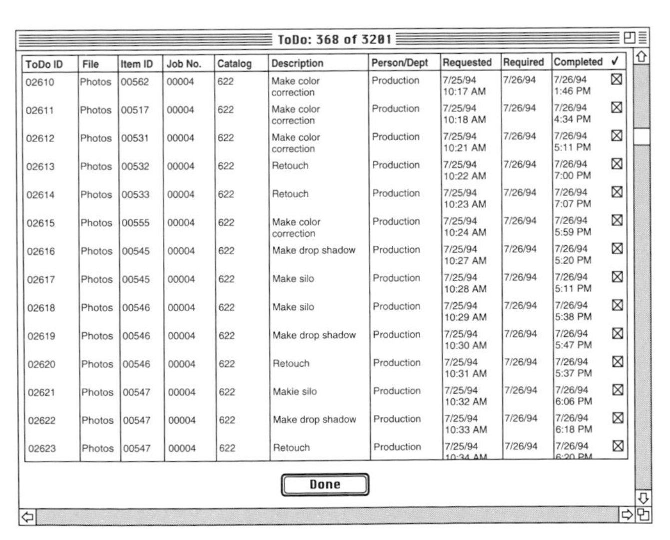

ToDo: 368 of 3201

ToDo ID	File	Item ID	Job No.	Catalog	Description	Person/Dept	Requested	Required	Completed	✓
02610	Photos	00562	00004	622	Make color correction	Production	7/25/94 10:17 AM	7/26/94	7/26/94 1:46 PM	⊠
02611	Photos	00517	00004	622	Make color correction	Production	7/25/94 10:18 AM	7/26/94	7/26/94 4:34 PM	⊠
02612	Photos	00531	00004	622	Make color correction	Production	7/25/94 10:21 AM	7/26/94	7/26/94 5:11 PM	⊠
02613	Photos	00532	00004	622	Retouch	Production	7/25/94 10:22 AM	7/26/94	7/26/94 7:00 PM	⊠
02614	Photos	00533	00004	622	Retouch	Production	7/25/94 10:23 AM	7/26/94	7/26/94 7:07 PM	⊠
02615	Photos	00555	00004	622	Make color correction	Production	7/25/94 10:24 AM	7/26/94	7/26/94 5:59 PM	⊠
02616	Photos	00545	00004	622	Make drop shadow	Production	7/25/94 10:27 AM	7/26/94	7/26/94 5:20 PM	⊠
02617	Photos	00545	00004	622	Make silo	Production	7/25/94 10:28 AM	7/26/94	7/26/94 5:11 PM	⊠
02618	Photos	00546	00004	622	Make silo	Production	7/25/94 10:29 AM	7/26/94	7/26/94 5:38 PM	⊠
02619	Photos	00546	00004	622	Make drop shadow	Production	7/25/94 10:30 AM	7/26/94	7/26/94 5:47 PM	⊠
02620	Photos	00546	00004	622	Retouch	Production	7/25/94 10:31 AM	7/26/94	7/26/94 5:37 PM	⊠
02621	Photos	00547	00004	622	Makie silo	Production	7/25/94 10:32 AM	7/26/94	7/26/94 6:06 PM	⊠
02622	Photos	00547	00004	622	Make drop shadow	Production	7/25/94 10:33 AM	7/26/94	7/26/94 6:18 PM	⊠
02623	Photos	00547	00004	622	Retouch	Production	7/25/94	7/26/94	7/26/94	⊠

Done

Assembly

West Marine uses catalogs in several ways. The master catalog is mailed and also distributed at retail sites. A wholesale catalog, based on the master, is distributed to marine product retailers with a different set of prices. Both catalogs include the full product line and are huge. As a result, admits Elwood, the catalog is more production-driven than design-driven. The first production task is to create a complete listing, with prices, of all the items offered by the merchandiser. The photographs exist to display the sometimes highly specialized nautical merchandise.

As Elwood explains, "Much of the photography is taken here at the catalog site. About 25% is sent from the original vendors [mostly 35 mm shots] and the rest we shoot ourselves. We bring in local professionals to our own in-house studio to shoot the small stuff. Then they go on location on the boats to shoot the rest. Most of the wearable boating gear [slickers, life vests] is modeled by West Marine employees."

On-staff writers generate copy to describe new products. Photographs are placed with the corresponding product description. Each page has 30 or 40 photos or illustrations, along with the complicated price lists and product descriptions. Design tasks therefore are extremely difficult to coordinate and especially problematic, even before the production stage.

The Old Production Cycle

When four prepress shops generated film, Elwood admits, "it was horrible, from a management point of view, to keep track of incoming and outgoing proofs. I've got a 3-inch binder from last year," he explains, "full of all the correspondence, memos, and schedules. Each shop had a different schedule, of course, and different billing rates and procedures. It was a nightmare. Just getting rid of the necessity of putting page files on a SyQuest disk, and assembling all the photos for each page, and sending them out every day to each of four different houses—getting rid of that alone was a huge step forward."

Each trade shop did high-res scans of the transparencies that accompanied the SyQuest cartridge. Then the shop would size the images to fit and place them on the pages, based on a low-res scan created at West Marine. The low-res scans caused problems, explains Elwood. "The trade shops would put pictures upside-down because they couldn't tell which way was up, especially on marine equipment where it wasn't obvious to anyone but a sailor. And maybe the comp wasn't quite obvious, since our scan resolution didn't make it clear which side was left or right. Sometimes the minor differences in cropping could be troublesome."

All these problems could have been solved by using true OPI, where any mistakes made in the design department could usually be corrected before they went to film. But OPI on such a massive scale is unusual in trade shops and made no sense unless a structured way of sending and tracking low-res scans by the hundreds were available. CatTrax was designed for just such a situation. There are now 6,000 images in West Marine's database.

Handing It Over

Elwood was pleased to hand over the purely technical tasks to the experts. "We are glad to let DCS be the experts on film production, trapping, and color correction. They take the time to create the drop-out masks and drop shadows. If we start trying to get into that area, we're not going to have time to do the creative work. I'd rather just have us concentrate on designing and laying out catalogs and let them do the rest of it."

Silhouettes, or drop-out masks, are particularly problematic in most production situations. Usually they are done after initial page layouts have been designed, since in most systems, silhouettes are difficult to reproduce accurately in an FPO (for position only) low-res scan and to translate exactly. In the

CatTrax solution, masks are made by the color-correction crew and placed in a database, along with drop shadows. This means that the designer has a nearly precise idea of how the silhouetted piece will look. This saves a lot of time on rework, according to Elwood.

Pickup is a lucrative part of creating most catalogs. In an average year, Elwood says, West Marine adds about 10% new products and drops about 5% of the old ones. (Last year, however, they added 400 new products, more than a 25% increase.) In a typical year, 85% of the old product images are picked up. All the secondary catalogs and mailers are 100% pickup material from the master catalog. Since the DCS images are on-line, easily accessed by part number, and kept available at all times, the frantic search for original photos and the need for rescans is minimized, resulting in a major savings of time and money.

Because of this, according to Elwood, once his group got the master catalog done, the little ones were actually very easy. "We just finished a 104-page mail-order catalog the other day, in fact, that was remarkably easy. It took two layout artists and one writer to crank it out in about two weeks."

Turnaround

Turnaround has been remarkable using CatTrax, reports Elwood. "We're getting spoiled here, you know. We finish a page, synchronize the files in a couple of hours over the modem, and then we get proofs in a few days. It used to be two or three weeks to get anything from a separator."

This capacity to turn work around greatly reduces production delays. According to Elwood, "This year, as it happened, we had some late pricing coming in and we had to get back with the printer and DCS and decide which press forms had to go where. We synchronized pages over the modem to DCS one night, and a couple of days later they were sending proofs to us and film to the printer. It all happened in a very short time frame when it finally came down to it. I was amazed."

Summary

CatTrax doesn't fit every site and probably cannot make every person happy. It's designed for catalog work and requires that you submit to the workflows invented by DCS. However, DCS has a very thorough understanding of the workflows needed by the catalog industry. The West Marine example shows that seemingly insurmountable problems can be solved by excellent planning and coordination. While DCS is working on adapting its tool set to a wider variety of prepress situations, so are other companies, anxious to provide solutions to the prepress industry.

CHAPTER 4

MAPPING WORKFLOWS

We have found that many sites have already developed workflow maps of sorts. They are usually linear and often contain "logic triggers"—IF/THEN/ELSE statements. The events occur in sequence and, theoretically, produce perfect film, perfect output, or a perfect print run. Unfortunately, these maps are seldom accurate. You may already have such a map for your own business.

A well-defined workflow map quantifies and structures the processes you must perform to achieve specific goals. There's no such thing as an all-encompassing workflow; rather, there are dozens, perhaps hundreds, of individual workflows that drive a project through your shop. Examples include: scanning workflows, trapping workflows, the generation of OPI (Open Prepress Interface) components and reference files, RIPping workflows, quality control processes, proofing, color review, archiving, deleting old files, and many others. Each process or subprocess requires its own map to direct you to the next one.

Productivity
You can't measure productivity in terms of throughput or how busy you are. Some shops we've seen track 100% productivity, yet their rework factor was over 30% —they were losing buckets of money. Fixing manufacturing errors and reworking files is not productive work.

We look on productivity as the ratio of billable vs. nonbillable hours. Some workflows are profitable; others aren't. There are workflows that cost you money every time you try them. Once you start to map each small process in your company, you will get a better idea of what type of work always runs smoothly and what type never comes fully under control.

Each new project usually has a different set of variables: the trim size, the types and quantity of images, screening requirements, imposition options, and a host of other project-specific attributes. You cannot predict those series of tasks you might have to perform in order to get a job done—particularly if you haven't done it before. In addition, there may be projects you're currently doing that aren't making a sufficient profit for the company. With thousands of possible production permutations, it becomes apparent that a single workflow map would fail as a management and process-control tool. Instead, mapping each process into a logical order of steps helps standardize jobs. If a particular process consistently breaks down, you can fix it. Without a workflow map, you'll have a hard time isolating and correcting problems.

The best mapping efforts we've seen generally break all processes into one of two categories: input or assembly. A workflow that creates new data files is an input workflow; a process that puts pieces together is an assembly workflow. When you mix and match input and assembly workflows strategically, they create a path that takes a project where it needs to go—usually, out the door.

Collecting Data
Automatic data collection is not trivial. Programs such as AppleScript allow savvy users to collect data automatically based on events (opening and closing files, starting applications, etc.). Some tracking applications build in this capability, so that an expert scripter is not needed. Tools similar to AppleScript do not currently exist on the PC or UNIX platforms.

Throughout this discussion we've said that workflow strategies are management strategies—they're not based solely on hardware, software, throughput, or technology. They're not based on "big iron" purchases (more and better equipment, that is), or whether you're using a Scitex or a Macintosh system; they're based on sound planning and the proper use of human resources. Five years ago it would have been difficult to map every process because it was difficult to collect data from non-automatic processes. That's not true today. Automated, digital technology now exists that addresses almost every possible production contingency—from page makeup to imposition.

The Five Markers

The Five Markers
There are five markers used to develop a workflow map. Each type identifies an event. Paths between the markers indicate the direction the events lead to depending on whether a decision is "yes" or "no".

Most geographic maps have visual markers to indicate the locations of streets, intersections, and landmarks. In a workflow map, there are equivalent "markers" that help you figure out where you are, and where you should go next. We've identified five markers you might find on a typical workflow map. (When we discuss charting tools toward the end of the chapter, we'll show examples of workflow maps using these markers that you might find useful in creating your own maps.)

Review

Review

Reviews assess information and check the accuracy of prior decisions. All workflows require reviews to ensure that a job is correctly executed and that quality standards are being met. Reviews are at the heart of any QC effort. They occur within input and assembly workflows. An input review happens when a project is first brought into your shop. Whoever puts the job onto the system reviews it to see if all the necessary information is there for the job to proceed. Reviews prompt questions about the information available or the situation. Is there enough disk space to accommodate the size of the files, for instance? A review within an assembly workflow might happen if an operator needs to use a previously scanned image from another project (a pickup). Are the images off-line? How will the operator get them—network, SyQuest, disk?

Decision

A decision is the answer to a question raised during a review and ultimately prompts the next step in the workflow. Examples might be: Are silhouettes needed? Yes. Are the fonts present on the disk? No. Are there any high-priority jobs to consider before we finalize today's schedules? Yes, the Harrison project needs to move to the top of the queue.

Process

Once a decision is made, someone or something initiates a process. Examples include: placing elements, printing pages, verifying OPI referencing, and loading off-line fonts from the site library. Many of the individual processes in the workflows we'll be looking at are repeatable and static enough to be candidates for automation (see Chapter 7, "Automating Workflows," for more information).

Transfer

A transfer happens when objects are moved from one "location" to another. Examples include transferring a file to the server, selecting Save or Save As, transfering low-res samples to disk or electronic postal site, printing the page (the file is moving from the monitor to the printer), and others.

Action

A required action is triggered and executed. Examples of actions within the workflow might be: contacting the customer service department to make sure there aren't any new higher-priority jobs that have the potential to "bump" other projects from the active sequence schedule; locating elements missing from the page, such as typefaces, vector drawings, or scans; or even calling the client for more information.

There are three logic triggers that occur when a review or decision is required:

YES The review and decision resulted in an affirmative response and the next event is triggered.

NO The review and decision resulted in a negative response. An alternative event is triggered.

RETURN A decision could not be made without revisiting a prior event. The job must go back "upstream," resulting in rework. (Always attempt to keep projects moving downstream. Any time work is pushed upstream, your profits are sinking. Rework isn't productive.)

Isolating Workflows

First, develop a list of every process that might possibly be required to complete a project. Think in small, manageable chunks. Remember the huge, all-encompassing linear maps we discussed earlier? They're an exercise in futility. Remember that workflow models are like reports. They're only useful if they help you process work or make decisions. The best way to map workflows is to create dozens of small workflows that can be linked together to produce an entire project.

We have found that assigning a code to each small workflow is also useful. Salespeople, CSRs, and estimators, for instance, can look at a master report and pick the codes and workflows that best suit the job at hand.

We're not going to map every workflow you might find in your shop. It wouldn't make sense. We don't know your business or the equipment mix you have. Instead, we'll do a few here for you to use as guides and let you try doing your own.

Let's start with two simple input workflows. The first is a four-color scan, with pleasing color and no requirements for color corrections or random proofs.

Code	Operation
1000	Review color
1500	Mount/dismount original
2000	Identify highlights and shadows
2200	Scan original
2500	Crop/name/scan image
2300	Save or Save As
2400	QC—scanning operator
5000	Return to CSR for routing

The first step is to review the original transparencies or prints when they arrive at the shop. (Review color is a step used in many different situations.) The review assumes you have good viewing conditions available, and that a member of the team knows what to look for and how to document casts, tints, or unbalanced neutrals. Other considerations are focus, clarity, brightness, and cleanliness of the original. We'll call this review Code 1000. After the review, the team makes the decision that no color correction is needed.

The next step or process is to mount the image. This code (1500) describes removing the last original from the drum or bed and putting the new one in place.

After mounting, highlight and shadow dots are identified (Code 2000). If Code 1000 indicated an acceptable original, this is all we really need to do to get a quality scan (assuming the scanner is capable and calibrated to balanced neutrals). If Code 1000 indicates that corrections are needed, the workflow would be different—it would include the necessary correction events to produce a quality scan (cast removal, pushing cyan into the grays, etc.). A workflow for images that need correction is different than a workflow for acceptable originals. The decision to use one code or another is determined during the original team-based analysis of the project. Using a list of possible workflows and their codes, the team can design a series of workflows that fit the assigned project.

Following the selection of diffuse highlight and shadow, the operator then crops and scans the file to a predetermined disk (Code 2500); a file name is assigned as part of the process.

Code 5000 routes the folder back to the CSR. This small, concise workflow produces a pleasing-color scan to be used downstream in an assembly workflow.

Here's another input workflow, listed by step, using an alternative coding structure. In this case an image is scanned for pleasing color, as in the workflow we just described; it also needs a Matchprint proof.

Code	Operation
0105	Review color
0100	Mount/dismount originals
0104	Identify highlights and shadows
0106	Scan original
0101	Crop/name/scan file
0102	Back up scan to optical disc
0107	Run random script
0108	Output random proofs
0109	QC—scanning operator
0501	Produce Matchprint
0525	QC—proofing operator
0105	Review color
5000	Return to CSR for routing
0900	Prep and check for shipping
0999	Ship to client

Script
A script is a small software program made up of high-level commands, written in a scripting language such as AppleScript. These scripts contain statements that perform tasks on the operating system level—opening, closing, moving, deleting, or renaming files; starting up or closing applications; writing messages to a file; and so on. These "programs" are easy to edit and test. They can be started whenever a certain event happens; for example, every time a file is placed in a certain folder.

Notice Codes 0109 and 0525—these are routine quality-control events invoked at appropriate locations in the flow. Remember, a QC event is no different than the scanning, stripping, proofing, or shipping events: QC must have the same—if not greater—priority as any other event.

Now let's look at an assembly workflow that merges, or joins, data files. It follows the same structure as an input workflow—with QC steps inserted after key steps.

In practice, projects generally require both kinds of workflows—input as well as assembly. Here's an example of a combination workflow:

Combination Workflow #96:
Assemble, trap, place hi-res, and generate Rainbow proof

Code	Operation
509	Preflight
542	Repair/prepare
838	Route for output
123	Electronically strip high-res
432	Trap images
853	Laser output
510	QC—position check
300	3M Rainbow proof
525	QC—proofing
009	Prep for shipping
200	Ship to client

Here's a similar combination workflow for scanning incoming transparencies and producing a proof:

Combination Workflow #96:
Scan, trap, place hi-res, and generate Matchprint proof

Code	Operation
434	Mount/dismount originals
542	Scan original
884	Crop/name/scan file
123	Route scans to assembly station
509	Preflight
542	Repair/prepare
123	Electronically strip high-res
432	Trap images
838	Route for output
334	Imagesetter output

572	QC—film
904	Matchprint
113	Review color
811	Route to CSR for review/routing
200	Ship to client

By developing combination workflows (composed of individual, process-specific workflows), you can categorize incoming projects and identify opportunities for parallel processing —that is, workflows that can be done simultaneously. With defined workflows at hand, you have a starting point for estimating time and materials, and developing an error-handling strategy, among other benefits. As you explore alternative technologies such as digital photography, direct-to-plate, and digital color, your established workflows provide a basis for new workflows. You won't have to start from scratch to determine if the purchase of a new device makes sense—you can measure its performance against the related workflows.

Once you've developed a comprehensive list of processes and code numbers required to achieve a finished product, mapping your workflows is greatly simplified. If you distribute the code book to everyone in the shop—from salespeople to shipping personnel—and solicit their feedback, you will begin to quantify and refine the map. Developing good maps is a continuous process, and one that allows for continuous improvement. If you don't know where you are right now and where you need to be in a year, then it isn't likely your profits will increase.

Matching Skill Sets to Workflows

So let's get started mapping workflows. First, pick a job that you feel is relatively profitable. Then take a blank piece of paper and divide it into two columns. In the left-hand column list the processes you think are necessary to produce the job. Once the list is finished, check it for completeness. Now, fill in the second column with the names of the individuals in your shop responsible for completing each task. Review the list. Is everyone qualified to complete their job? Do they have the skills and training needed to fulfill their responsibilities?

Next, erase the names and replace them with skill sets. This exercise lets you see what needs to be done to get a job in and out the door most effectively. Does your staff have the prerequisite skills? (Notice that there may not be a single department manager anywhere on the sheet.)

Repeat this process for twenty more projects. Note which ones give you the most trouble—you know, the jobs that always require rework. Consider too that not all workflows are profitable and certain clients might never be profitable. There are good clients and bad ones—make sure you fill your book with the former and discourage your salespeople from bringing in the latter.

Once you've completed this exercise you will be able to identify which disciplines you need on your production teams. Normally the list includes a salesperson, a CSR, an estimator, a person from purchasing, an experienced systems operator, a retoucher, and perhaps a stripper. If you have people in your organization who never find their way onto a production team list, either retrain them or eliminate them from the payroll. Otherwise, you may never be as profitable as you could be.

When the list is complete, review it closely to ensure it represents how you want jobs to be done and what skills are required to do them. Match the right people to the right tasks and see how productive they can be. Imagine hiring new employees with no idea of how your shop runs—your workflows provide specific instructions for making sure tasks, and ultimately jobs, are fulfilled to customer specifications and satisfaction. In the workflow examples we've shown, scanning (for instance) is a single code. Is that accurate for your shop? What if your shop has flatbed scanners for line art, drum scanners for high-quality color, and a digital camera for product shots? Each of these input options needs its own workflow. It also requires the right level of experience to perform the task. You might, for instance, have less experienced employees running the flatbed scanner than you do operating the drum scanner. It's certainly more cost effective than paying your most skilled scanning operators to do flatbed work. Naturally, each site is different, and has different considerations to make.

Tools for Mapping Workflows

Now that you have identified all the processes and subprocesses that make up your daily routine, and have matched skill sets with the work, you can create a graphic representation of the workflows that suit your business.

There are many fine developers working on software products designed to assist in the workflow process. Many share common attributes, and most are designed to interface with accounting or estimating. Since businesses vary, the package you ultimately decide to use must be weighed against your specific needs. Before you purchase, consider the burden of maintaining the software, its complexity, the required learning curve, and the reports it's able to provide. In our opinion, reports are the most important consideration of any workflow-related software program. There is, however, one exception to this: mapping or flow-charting software.

Flow charts are easy-to-understand, visual representations that help map, implement, analyze, and refine individual workflows. There are many flow-chart packages available. For our work, we use a program called MacFlow from Mainstay software (591-A Constitution Avenue, Camarillo, CA 93012; phone: 805-484-9400; e-mail: Mainstay@aol.com). This program is also available on the Windows platform and is called WinFlow. While there are probably many alternatives, we find

MacFlow elegant, simple to use, and highly effective. Purchasing such a program will save you a lot of time in your mapping efforts—and provide you with maps that can be easily updated for distribution to the responsible workflow teams.

A good mapping program helps you manage *granularity.* We use this term to indicate the level of detail you intend to map. Granularity is best described by defining its two extremes.

Here's a workflow map with very little granularity:

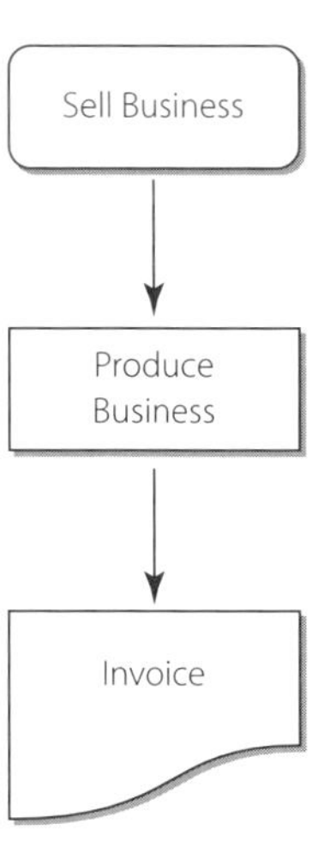

For those readers that find this amusing, this map accurately depicts a business that doesn't use mapping strategies.

Clearly we need more granularity to work effectively. Overzealous detail, however, can be too much of a good thing.

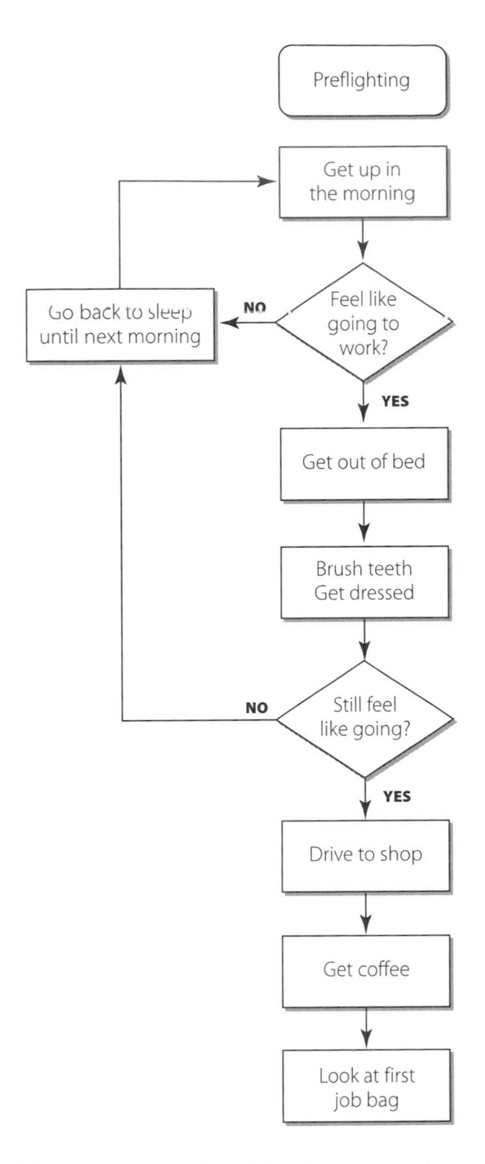

There is a relationship between the overly simple map and the overly complex map: the highly detailed map is actually a subprocess of the simplified map. But who would track such useless information? Nor does the "Sell, Produce, Invoice" model serve any purpose. What we need is a map that lies somewhere in between these two extremes.

A good flow-chart example is preflighting. In simple terms,

there are just four steps: receive the job, check it for completeness, create a report listing what's wrong with the file (so someone can fix it), and route the job to file repair. This is one possible preflighting workflow, where salespeople and customer service reps, with limited access to equipment, do a check for completeness. In another workflow, they might be responsible for actually repairing some of the problems they find.

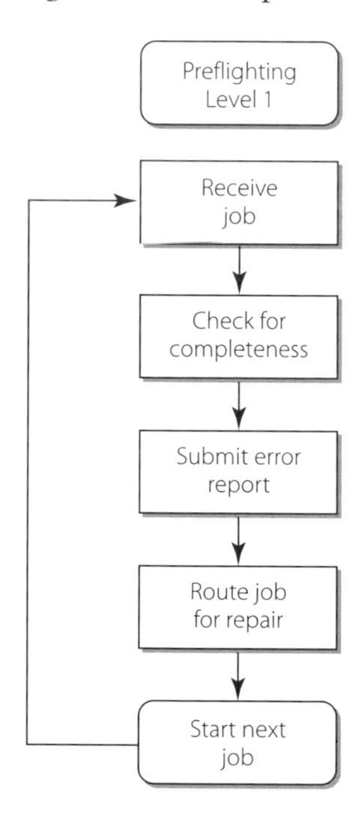

Let's start with "Receive job." A charting program will normally allow you to double-click on a task symbol and link it to another map—which shows the subprocesses associated with that task.

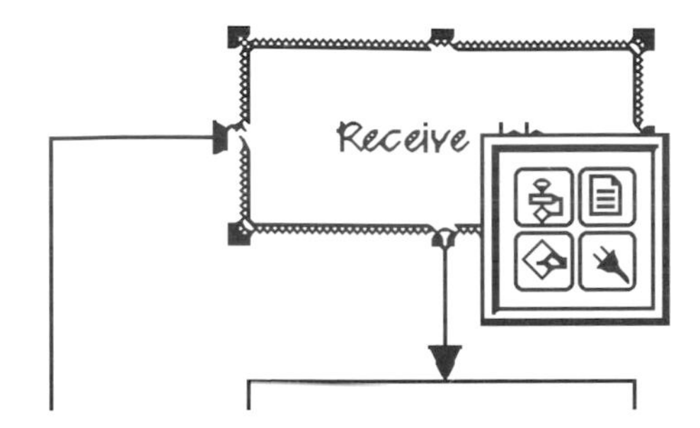

Now we have a blank screen and can build a map for the subprocess.

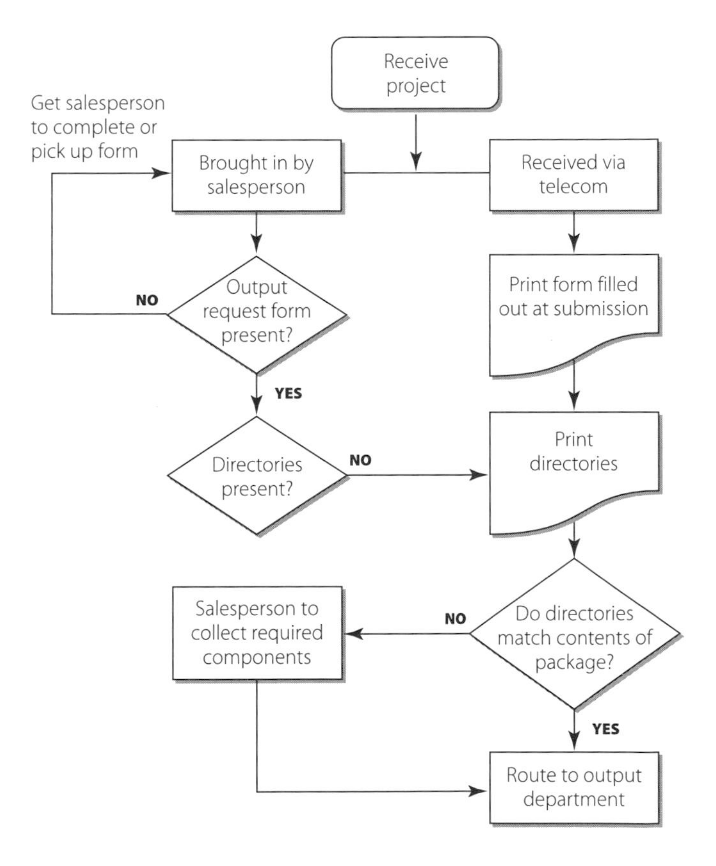

Back in the primary chart, a small marker indicates that the symbol contains details of a subprocess.

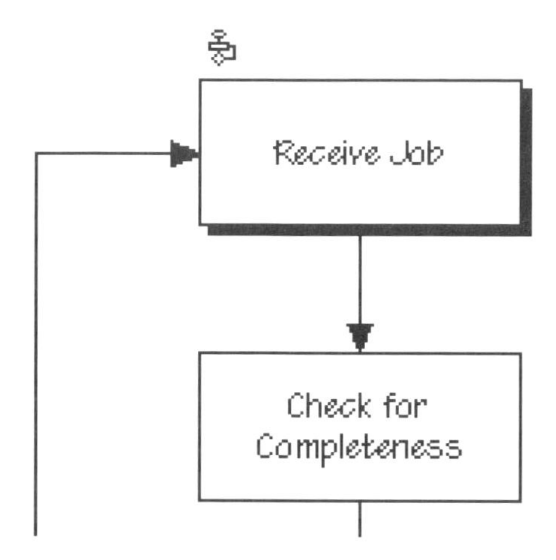

Some flow-chart software applications allow you to add notes, sounds, and even videos to a chart.

Using a charting tool to draw your maps can be very advantageous. First, an entire library of different workflows can be saved and coded in such a manner as to link them back to task codes used by estimating, customer service, and manufacturing.

Charts can be distributed electronically (MacFlow can create "stand-alone" charts that don't require the application to view) or they can be bound simply into a three-ring binder (the best alternative when you're first implementing a mapping system).

The very process of creating flow-charts forces you and your staff to think hard about how projects are accomplished. Most important, once you've defined the most effective and profitable workflows, they can be applied to new business—whether from an existing or prospective client.

Some companies we've seen have developed icons to use in place of markers, such as printer icons, disk drives, workstations, output devices, people, and other visual signposts. The extent to which you play with your maps is up to you. Don't forget, however, that the actual mapping function isn't productive; you can't bill anyone for the time and effort you invest in making your charts. It's a preparatory effort designed to provide you

Simple Input Workflow

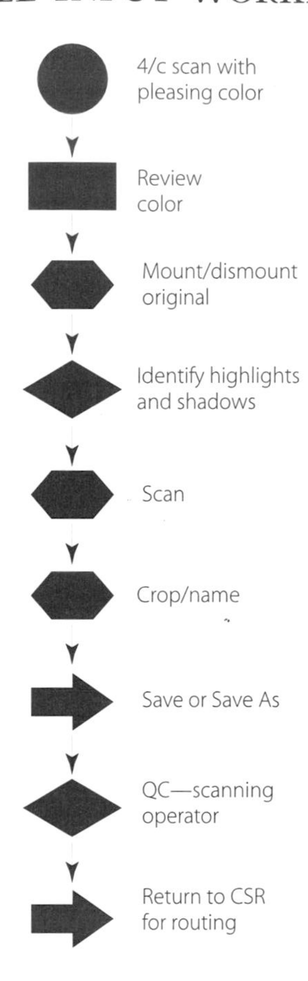

Scanning Workflow: No Proof

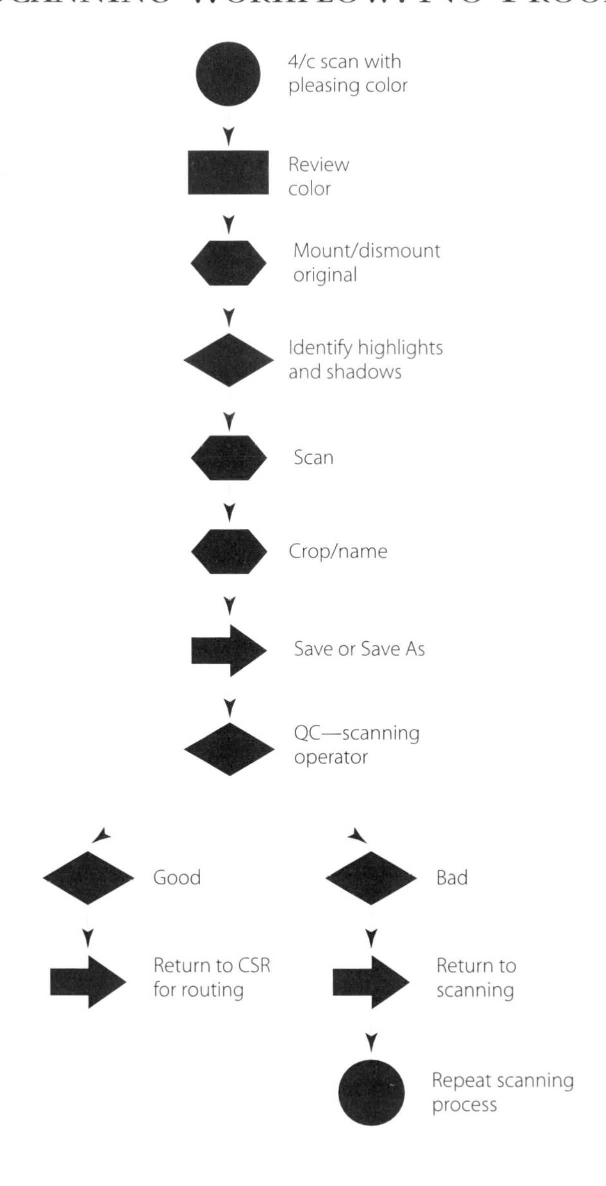

Scanning Workflow: With Proof

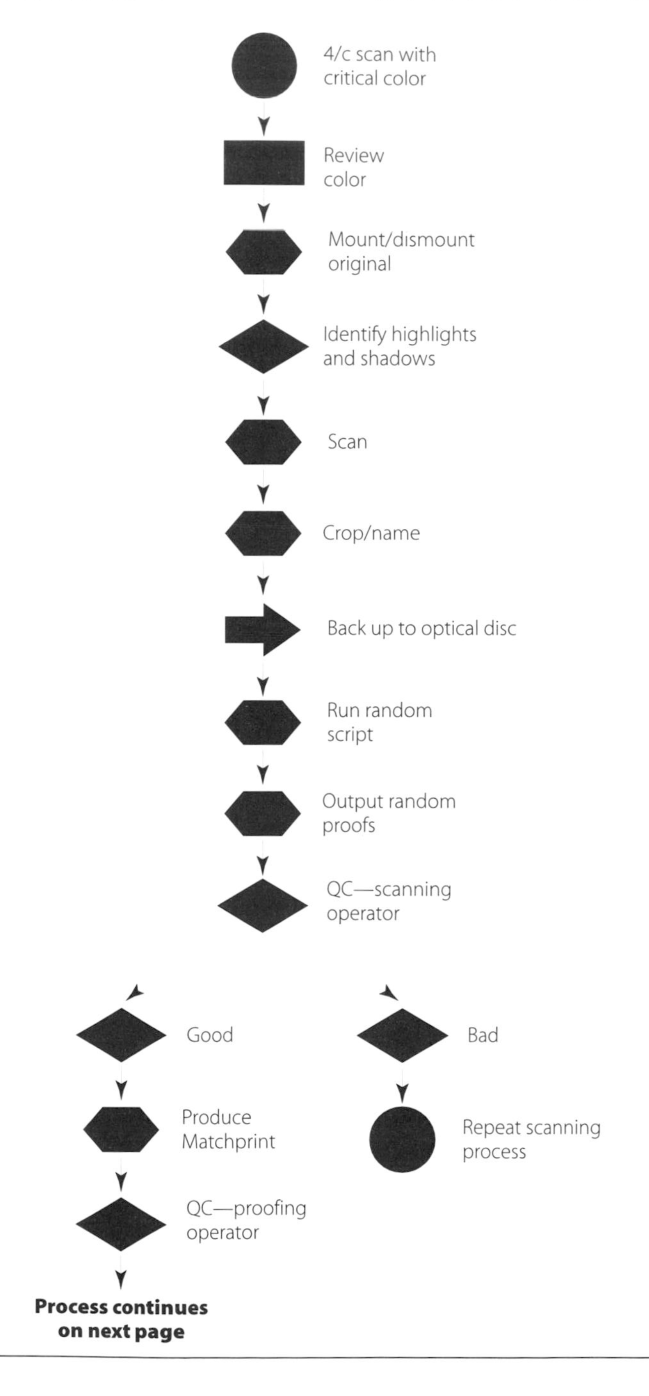

Scanning Workflow: With Proof (continued)

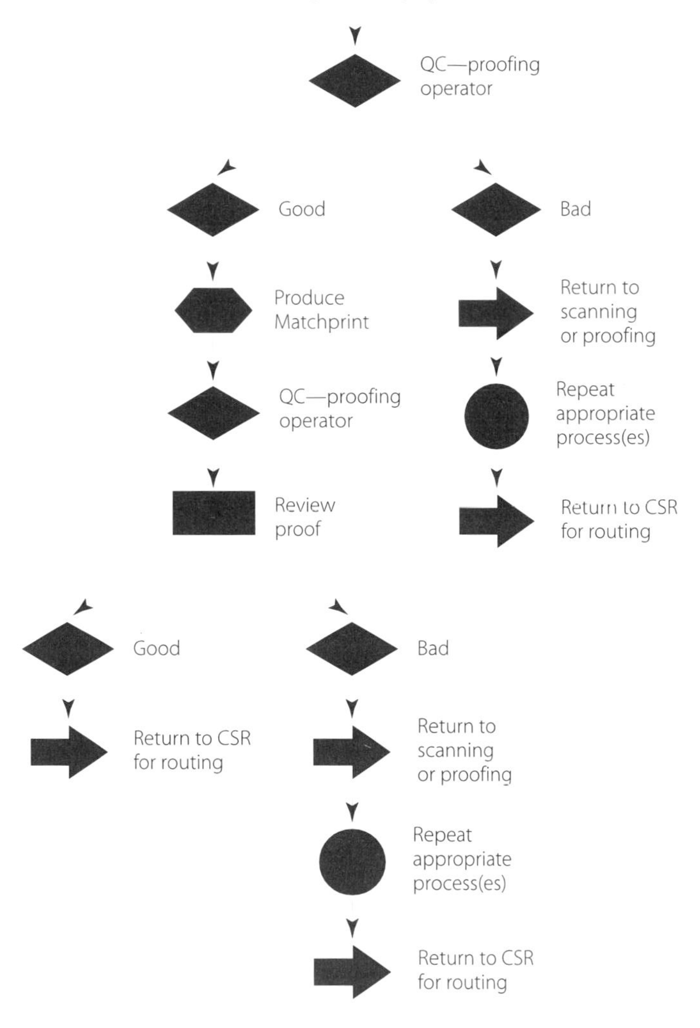

CHAPTER 5

Strategic Approaches: Process Improvement Points

Once you've defined and mapped the various workflows within your shop, the next logical step is to identify areas where you might make improvements. Improvement—at least within the scope of this book—means increasing profit margins.

There are two key areas to focus your improvement strategies. The first is productivity. How do you define productivity? Whether you're thinking about people or equipment, the definition remains the same. Productivity is reflected in hours that generate income; nonproductive time (and there will always be some) is time or resources spent that don't generate income. The starting point for improvement, therefore, should be defining the maximum percentage of productive time you could achieve under perfect conditions. A successful strategic plan could be built on the assumption that 5% of your shop's time will be allocated to staff training, administration, management, and accounting. That leaves 95% of all available resources—time, equipment, and capital—to be productive. If you had to guess, where would you place your current productivity level?

Let's look at a typical 40-hour shift. If your productivity level is at 95%, 38 hours of each shift are generating income. At 80% the figure drops to 32 hours making money and 8 hours lost to inefficiency. In our experience, many shops fall well below the 80% level—prior to implementing an improvement strategy. It's not surprising that improving productivity consistently results in increased profits.

The second area to focus improvement is to increase the percentage of first-pass client approval. When digital technology was first used by designers, many producers had a difficult time changing their processes to accommodate the new project "model." Cultural resistance aside, the old production processes were simply too rigid to adapt to electronic mechanicals being dropped into the existing manufacturing cycle. Now that the technology is well accepted, and the processes are beginning to stabilize, the importance of a clearly defined workflow strategy should become clear to you.

To help you develop a strategic approach for improved profits, we've created a list of Process Improvement Points (PIPs). We have tried to structure each PIP consistently to include: an Abstract, a Strategic Approach, Financial Considerations, Resources Required, and our own Observations. Naturally, each site is very different and the effectiveness of each PIP can only be determined by thinking about them in terms of your own business. The list is based on personal experience working in the field and from the extensive site visits we routinely perform.

We have only one guideline on how to use this list, and that's to prioritize the PIPs according to their worth to you. Measure each carefully and consider what the financial impact on your business will be. Try implementing the ones worth the most and leave the more abstract and esoteric ones (as defined by your specific environment) for last.

PIP
1

Training Clients Upstream

ABSTRACT

The jobs you process start at the creative site—and as such begin their lives outside your control. Decisions made by creators have a direct—and often negative—impact on how profitably their work can be processed.

STRATEGIC APPROACH

Training your clients can have several positive results: an immediate improvement in the stability and correctness of incoming files, a greatly improved ability to realistically estimate projects, improved throughput, and a better understanding by the designer about how a project unfolds when it gets to your site.

Since training clients requires that you understand how they process their jobs, your staff will develop a much more intimate and proactive knowledge of the workflow map at the creative site—which is, after all, the source of all your work. We're not talking about PageMaker 101 either. As a vendor, you need to provide workflow-specific training with one thing in mind—improving profits at your site and making it easier for the client to use your services.

FINANCIAL CONSIDERATIONS

Costs for an organized training effort depend primarily on two things. First, do you wish to provide a classroom environment? Studies show that organized training with an experienced teacher and effective materials, delivered in a classroom environment, are more effective than videotapes, self-paced materials, or one-on-one on-site training. A simple yet formal environment can be designed with two or three student workstations, tables, a blackboard, and an overhead projector. We have found that the more formal and organized the training effort, the more effective it is. The client, upon viewing a formal classroom, is more likely to associate a monetary value with the training you offer.

The second item to consider is your cost to maintain training personnel. Will your instructors be full-time or will they be split between training and other functions? Third, you must purchase—or develop—course materials. You cannot train a class full of adults without a lesson plan. The more structured, organized, and (most important) relevant to your specific creative/production workflow the lesson plan is, the more effective and well-received it will be.

Resource Considerations

Videotapes and self-paced instruction aside, providing formal training requires a classroom, a teacher, and courseware. Trainers represent the largest ongoing expense of all three and fall into one of two categories: they are either fully dedicated to the training function, or it represents only a fraction of their time. The full-time trainer is one whose duties are solely related to the training function. This individual's focus should always be training, but he or she should also be a spokesperson for training during major planning efforts. In this way, the trainer can contribute expertise to help determine what level of training is needed, how to best deliver the required instruction, and how much the prescribed delivery will cost the producer for any situation, whether training is required for new equipment, new employees, new workflows, or new software.

Don't mistake your system-savvy experts with professional trainers. No matter how good someone might be at putting out fires, they might not be great trainers—and they often don't have access to the resources that a real trainer needs to be successful. Filling both needs (the fireman and the trainer) with a single individual is a difficult—but not impossible—challenge. It just takes a special kind of person to wear both hats. And for your part, you must provide the fireman and the teacher with two complete sets of resources.

Observations

Graphic arts producers pay for a lack of structure in how their clients construct creative files. Design and creative companies

Bottlenecking Automated Imposition Efforts
We have found that many of the problems with automated trapping and imposition programs occur because of file irregularities. These are not actual errors on the part of the designer, but reflect some of the peculiarities of saving files in EPS format. Simply learning how to construct files that don't have any hidden "gotchas," such as multiple layers of object grouping, can make a major difference in the ability of these programs to process files.

pay as well, through their own lack of productivity and tighter-than-necessary deadlines. Upon analysis, the same types of problems continue to crop up. If you closely follow the evolution of a typical project, you will quickly identify the point in the creative process where font problems start; where the trim and bleed is set in such a way as to bottleneck automated imposition efforts; and where traps are incorrectly assigned or inadvertently enlarged or reduced when artwork is manipulated in a page layout program. Most of the problems that make your work slow down, or stop, begin upstream at the creative site.

When designing a training program, there are several issues to consider. First, will you charge your clients for training, or will they (and your own sales force) feel that they should get it for free because they already spend money on their printing, scans, or other services? There is good reason to believe that when something is free, the recipient fails to understand or recognize its true value. If you charge the client for training—or at the very least clearly assign a stated monetary value before you give it away—chances are, they will respect it, and you, more.

In any case, formal training carries a considerable cost burden. It must be paid for—whether budgeted, expensed, and reflected as an internal operating cost; or borne by the client. You should view the depreciation of the classroom, the teacher's prorated salary, and any courseware expenses as an account category in your financial statements. Even if training fails to make money, it can at least pay for itself. Training is valuable to the student and we feel strongly that you can charge for it.

Another serious consideration is that if you do charge for your training (to offset the costs, establish value, or actually generate a profit), it must be comparable to other learning products. When you begin to offer structured commercial training you enter an arena where dedicated professionals are competing for those training dollars. You must consider the look, structure, and coverage of your courseware materials; the experience level of your training staff (as it relates to effectively delivering adult education); and the quality and inventory of your classroom. In short—if you decide to train your clients, set standards as high as those of your core products. Don't forget, your competition isn't only from professional training organizations—it's also coming from other service providers trying to gain a competitive advantage over your organization.

PIP 2

Staff Training

Abstract

Training your internal staff is at least as important as training your clients. You have many different workflows that require specialized skills. Identify these skills and provide the training that each individual must have to get their job done perfectly.

Strategic Approach

There are several advantages to a structured training program for your staff. First—and most important—is having people who are thoroughly competent in the technical responsibilities that their job description calls for. A perfect example is the customer service representative, who in many cases is expected to preflight electronic mechanicals with no knowledge of basic system or software functionality, and without access to a machine on which to check incoming projects.

Financial Considerations

Costs for internal training include: the instructor, the classroom, and the training materials to be delivered. Every time a staff member enters the classroom, there is a cost incurred. These costs must be viewed as a budget item. The return can be measured by a reduction in jobs that either slow down or stop the processing of work in the shop, thus improving productivity.

If client training and staff training are viewed as separate, but similar, components of the same strategy, management can

Software Upgrades
The differences between, say, Photoshop 2.x and Photoshop 3.0 is likely to affect virtually every job coming into the shop, yet it is assumed that everyone will learn by osmosis what the new features are, what bugs have been fixed, and what tools have been extended. Sure, the more ambitious workers will learn something, but others will only half-use or half-understand the new features, making them work less efficiently.

benefit from economies of scale. The instructor responsible for client training can (and should) coordinate the training materials, delivery methods, and environment to meet staff training as well. A classroom engineered for client training should be used to provide training to the staff working with those clients. By coordinating the training program between client and producer, the instructor can work in concert with manufacturing team leaders to tailor training to each specific client.

There is a natural fear that taking people away from their normal duties reduces productivity—and you're right, it does. At a 95% productivity level, however, the 5% left over should give you more than enough opportunity for training your staff.

Also, be aware that software and/or hardware upgrades usually require additional training. The real cost of an upgrade must take into consideration the cost of teaching everybody new features or functions. Don't make the mistake that many managers do and simply distribute the upgrade without paying attention to this critical training issue.

Resource Considerations

You can either feed your staff knowledge a little bite at a time, or at regular intervals. Frankly, we think that three meals a day are more nutritionally sound than eating on the run. Training is an ongoing process that parallels changes and growth in your organization. The best option is a full-time trainer equipped with highly workflow-specific courseware, a well-designed classroom, and a schedule that is well-balanced in its approach to developing skills in your staff. Clearly, though, this option is outside the reach of many—perhaps the majority—of small- to mid-sized shops. Whether you're a 5, 50, or 500-million-dollar producer, training must be structured, organized, and ongoing. The only difference between your training program and someone else's should be measured in terms of staffing and scope—not approach or intention.

Observations

Within the producer environment there are often glaring deficiencies in staff training. The most important place for you to look is on the workflow map. Since each process on a workflow map requires either a) a human being making decisions and/or affecting data in some way, or b) equipment and/or software performing an unattended task, it is a logical jump to determine what skills each individual must possess to best achieve their task(s).

Just because someone knows a technical subject, you can't assume they know how to *teach* that subject to someone else. Often, exactly the opposite is true—techies can make terrible instructors. The larger and more formalized your training effort becomes, the more you should consider training your *trainers* in issues affecting adult education. Adults learn in very specialized and predictable ways—ways not related to the subject matter. Neither are they related to the experience or technical expertise of the instructor. Many technical people who also provide training to others fail to realize this and might even have a hard time admitting it. Do you have any way to test the effectiveness of your teachers? (Formal certification programs are beginning to appear. GATF and Adobe Systems both offer such programs.)

GATF
The Graphic Arts Technical Foundation in Pittsburgh, Pennsylvania, researches technical issues and develops standards in the graphic arts field. The consortium's members include a wide variety of vendors, graphic arts firms, and universities.

Another issue is that of perceived and real value to your employees. Training is a valuable and costly commodity and must be treated as such. Some managers go so far as to insist that employees apply for training programs—as opposed to simply telling workers that the class starts on Monday. One option that we've seen in the field is having the student sign an agreement that spells out financial penalties to be imposed should the employee leave the company within a period of time following the training. Based on a sliding scale, these penalties reduce (but by no means eliminate) the danger of an employee being trained and immediately snatched away by a competitor with a better offer.

PIP 3

System Integration Upstream

ABSTRACT

System integration is the process of designing—and perhaps even selling, installing, and maintaining—a graphics network. The systems being used by your clients can have a negative effect on your ability to profitably process their jobs. You might consider assuming varying degrees of responsibility for how their machines are configured. The more they comply with your manufacturing methods, the more money you can make processing their work.

STRATEGIC APPROACH

By mandating software versions, hardware choices, and how they are being used, the vendor, capable of making the investment in time and money, can gain a degree of control over how its client's projects are being assembled that is tough to achieve under any other circumstances. This option is particularly well-suited to environments where creative, production, manufacturing, computer purchasing, networking, and upgrade policies are already somewhat centralized, such as a newspaper, a catalog environment, or any other situation where multiple disciplines are controlled or affected by centralized management.

The more stable and predictable the client's job submissions, the more effectively you can design workflows that maximize your particular mix of people and equipment. After lack of training, the largest contributing factor to workflow "stops" is the system environment on which the clients created their files. This relates to software versions and even hardware, such as scanners, monitors, output devices, and other factors.

FINANCIAL CONSIDERATIONS

As a production site, you can begin a system integration effort simply by putting a color output device—of your choice and under your control—at your client's site. One of the World Color sites we visited, for example, provides specific catalog clients with a 3M Color Proofer. By having it generate the color lookup tables used by the device (based on output characteristics at their site), they have been able to remote-proof pages for position and even some color via an ISDN line that they also purchased and installed. The up-front expenses pale in comparison to the savings realized by World Color.

Resource Considerations

It's likely that you've already dealt with many hardware, software, and connectivity issues. The individuals within your organization who are currently responsible for technical product testing, evaluation, and selections are the same individuals who need to be part of a team deciding whether or not to supply a creator with specific hardware or software. A well-thought-out approach would have your hardware and software gurus hammering out specific client integration strategies with manufacturing and sales. From there, it's simply a matter of accounting for the costs: does the account have the room to cover the cost of the proposed solution, or will you simply charge the client directly?

You might consider simply charging the client for the proposal and recommendations that you make. A solid proposal provides an executive overview, a table of contents, an introduction, discussions about all aspects of the installation (training, equipment required, networking considerations, etc.), and the costs associated with each stage. Many dedicated system integrators charge for their proposals and deduct the cost of preparation if it's ultimately accepted.

Observations

Lacking the experience of having done many installations, well-intentioned individuals often underestimate the cost of ongoing maintenance of software and system upgrades. A professional integrator allows for these costs and also considers the appropriate use of the components and training users how to best use them. The integrator and trainer should communicate with manufacturing to ensure that system design, training, and manufacturing requirements are all addressed.

A survey of the top 20 largest service providers would show that most of them have (at some point) assumed the role of system integrator—going so far in some cases as to provide equipment and training "free" in return for a printing, color, or output contract. Simply a form of up-front rebate or discount, this tactic often benefits both creator and producer. The manufacturing site secures a steady supply of work, gains control over the client's equipment, and effectively shuts out competitors for the length of the agreement. The creator receives a considerable amount of capital equipment, some amount of training, and (theoretically, at least) improved communications with high-end systems, which results in the vendor being able to produce within a tighter schedule (because there are no surprises), and both sides maintain fixed costs to design and produce digital pages.

While offering certain advantages, adopting the role of integrator has its share of problems. The costs are considerable, as is the drain on human resources—support can become a major expense (and headache). Another danger is that everyone's an expert—and everyone has a different opinion as to what combination of hardware and software is best for specific conditions. Your credibility can easily be questioned by someone (such as another integrator) who has little concept of exactly why you recommended a specific component. Tread carefully.

PIP 4

The Salesperson as a Member of the Production Team

ABSTRACT

Do you pay commissions to your salespeople? In our opinion, compensation must be tied directly to profitability. And profits are the result of finding work that best suits your workflows. Salespeople are tempted to bring in any work they find—whether it fits or not. By helping your salespeople understand which workflows are profitable and which ones aren't, you will begin to see your profit margins improve. You can only achieve this by increasing your sales staff's exposure to manufacturing and workflow strategies.

Strategic Approach

Salespeople
The nature of graphic arts sales is changing radically. Buyers are more concerned about technical issues, so salespeople must be prepared to answer complex technical questions. And buyers are more often women, who are less amenable to the good-old-boy sales approach.

Salespeople in our industry have largely been left out of the digital loop. The clients using desktop systems to construct their pages understand (hopefully) how their equipment functions and how each project is built. A salesperson that doesn't understand how that work will be processed, and how profitably it can be produced (given your specific workflows), is at a major disadvantage. A salesperson could easily think that any job produced on a desktop system could logically be processed on yours—this is not always so. Make sure your salespeople know how to match prospective work to proven workflows. Not all work is good work, and not all clients make you money. You can't afford to lose money on every invoice and make up your losses through volume. Volume has little to do with profits. While many factors impact profits negatively, few are as negative as a poorly focused sales force.

Financial Considerations

All jobs look the same to many salespeople—money in their pockets. They must be made to realize that the market sets prices—your efficiency level determines profit. One shop can lose money on work that's profitable somewhere else. Profit is directly related to the fit between the processes and work profiles you do best.

It is management's responsibility to impress on their sales staff that time spent learning which workflows are profitable will increase their income, not distract them from their appointed rounds. Here's a twist on an old saying: "If you don't have time to do it right the first time, how can you find time to fix it?" Support and rework is also directly related to only selling work that makes you money—and passing on work that doesn't.

Resource Considerations

Workflows must be defined and then discussed—repeatedly—with salespeople. You probably can't teach a salesperson every technical issue that might crop up. In some cases project-specific technical issues could spell the difference between acceptable profits and low margins. You will occasionally need to send technical people with the salesperson to act in a sales-support capacity. Your technical staff can provide insight into specific workflows at the client site and understand how they affect corresponding processes at yours.

Be careful that your salespeople don't abuse your support people, though. It's easy to fall back on bringing a techie to every sales call. This can be avoided by defining your workflows in an easy-to-understand manner and continually working to make them clear and concise.

Observations

First, define all the profitable workflows that match your specific conditions. Describe them clearly and create a simple-to-understand workflow map for each. Provide these to your sales force and make sure they study them.

The more each employee knows about the duties and responsibilities of other individuals in the organization, the more successful a strategic approach to process control is likely to be. Many salespeople—in particular, really high-powered, high-earning salespeople—find it very easy to resist any change in their job requirements. This makes management of these same individuals difficult. If the salesperson sees team production meetings as a waste of time, they are likely to resist—leaving you with few pleasant options to force their participation.

As your company evolves into a workflow process, it will increasingly be dealing with new production methods such as direct-to-plate and digital color printing. A workflow that results in profitable page output under these conditions is as different from the workflow for offset as it would be for multimedia distribution.

Discard workflows that cost time and money and produce less-than-acceptable margins. Identify profitable workflows and sell into them.

PIP 5

Intracompany Communication

ABSTRACT

How are you organized for effective communication? The more communication occurs between members of your staff, the more likely they are to come up with ideas to improve profits and ease workflows.

STRATEGIC APPROACH

How many departments does your organization maintain, and how many employees are in each department? If you subscribe to the team concept, each individual of a department becomes a member of a cross-discipline production team—ensuring that they understand the needs of others in the workflow. How much do they interact with each other? Several innovative managers have designed incentive plans with the express purpose of getting staff members thinking about how to make their particular processes more effective and themselves more productive.

FINANCIAL CONSIDERATIONS

Each department is encouraged to develop programs that will either save money, time, reduce bottlenecks (and latency) and generally speed work more effectively through their department. One program idea is to establish a point system of some sort, where credits are awarded to individuals within the department for ideas that meet a set criteria. These credits can be for additional paid vacation (seemingly the most popular among staff members), money (also very popular), or, as is the case at several sites, hard goods such as short trips and outings, bicycles, sporting equipment, and similar "fun" products that the employee might love to have but could not normally afford. We know of one site that goes so far as to print a black-and-white catalog of items available through the program.

Resource Considerations

If your company is rife with middle management, this is an opportunity to get them out of their meetings and into something productive—such as designing and implementing a reward program—that encourages peer-to-peer interaction. The basis of any reward should be directly related to the amount of money saved by the idea. You would be absolutely amazed at the depth and breadth of interaction this system will invoke.

Observations

We can make broad recommendations about ways to foster good communication within and among departments. Your employees, on the other hand, know your specific business at least as well as you do—certainly within the context of their own responsibilities or those of their departments. Use their input—good ideas abound, but there often isn't any place for employees to channel them. Provide the outlet, and the ideas will fly. You will actually begin to see competition among departments for who won the most points, who had the best ideas, and which department is having the most powerful impact on the company. Competition like this is good for your organization—great ideas that fall on the deaf ears of uncaring management can cause resentment and actually suppress thinking.

Think about it for a minute—how would you like every single person in your entire company spending at least part of their time trying to make the company more productive?

PIP

6

The Role of the Customer Service Representative

ABSTRACT

Perhaps the most critical position in the digital workflow is that of the customer service representative (CSR). Before the emergence of digital pages, the customer service representative worked in a world where all projects were composed of visible elements. The CSR was able to physically assess the completeness of a project and route it accordingly. If the tissue overlay presented specifications for a specific element, and the element wasn't physically in the job bag, the CSR was able to act on the situation immediately. CSRs could often solve a problem with little or no intervention from manufacturing personnel, working directly and independently with the salesperson and the client, or the client alone.

STRATEGIC APPROACH

Educating and nurturing CSRs is a very important area for you to focus on. The challenge is to define the job and then match the skills (both technical and communication) and personalities to that job description.

What is a CSR? In some cases they share responsibility with a salesperson. Their duty is normally to follow a project through the factory and make sure that all the components are in order at the specific moment in time that they're required.

Many owners and managers consider this position to be somewhat of a drudge job best filled by low-cost employees. There's considerable turnover in this area throughout our industry. Why? Because the job is thankless. Everyone—the client, the salesperson and the manufacturing personnel—blames the CSR for no reason other than the fact that they're sitting in the middle of all the forces involved. Something missing? Yell at the CSR. Font substitution? Yell at the CSR and tell them to call the client immediately. Water cooler out of H_2O? Yell at the CSR. In fact, yell at the CSR for the hell of it.

FINANCIAL CONSIDERATIONS

The truth is, you really can't afford turnover or dissatisfaction in the CSR department. Instead, plan on spending more for quality people and provide the training and equipment they need to do what you're paying them to do.

To redefine the role of CSRs, first determine how they should preflight your incoming work. Don't let work move past the CSR's desk without at least a cursory analysis of the job's completeness. The cost of addressing this issue includes the cost of training the CSRs and providing them with an available workstation on which to check the contents of incoming disks (or to open files if they're responsible for Level 2 preflighting; for more information about Level 1 and Level 2 preflighting, see pages 115–120).

The longer it takes to find a problem, the more it costs to fix it. This basic fact of life was greatly exacerbated when digital pages were introduced to the producer environment.

RESOURCE CONSIDERATIONS

The best approach to reengineering your preflight and customer service processes is to create a single interdisciplinary team. It would include a willing client, the salesperson, a CSR, a technical support person, and any other individuals you think are vital to the process. Once this first team works out the bugs and shows an increase in productivity or acceptance rates, use them as a model to train other CSRs and their teams.

OBSERVATIONS

Consider the physical location of the customer service department. Is it on the shop floor mixed in with manufacturing, or is it hidden behind walls and closed doors somewhere in the bowels of the building? The closer it is to where the work is being done, the better it works. Ideally, it should be smack in the middle. This architectural consideration improves communication and reduces contention among manufacturing and customer service. It also brings salespeople into closer contact with the work they have in the factory during their meetings with their CSRs.

In a reengineered environment there is a difference in the roles and responsibilities of certain employees compared to a traditional shop. Nowhere is that difference so apparent as when you analyze the responsibilities of the CSRs. Their roles really haven't changed so much—they still act as the liaison between the factory and the client, or the factory and the salesperson. In terms of CSR responsibilities however, they have changed dramatically. Unfortunately, even if their only responsibility is ensuring the completeness of a project, they rarely possess more than a rudimentary idea of what to look for. Without a structured workflow for receiving digital work, they lack the direction they need to do their jobs. A structured receiving process would include forms filled out by the client and/or salesperson, and client-supplied laser output.

PIP 7

Estimating Costs and Ongoing Cost Feedback

ABSTRACT

Before a manufacturer begins a job, an estimate of how much the job will cost to produce is made, and roughly how much profit it should generate. Very few sites have a structured way to recheck these estimates following production of the work. This is a serious impediment to profitable operations.

STRATEGIC APPROACH

Comparison of actual and budgeted expenses is probably a fundamental building block for your company's normal accounting functions. Do you have a structured program for revisiting

Curve Fit
This is a technique many of us learned in high-school chemistry class and one that marketing managers creating new products, and economic forecasters, specialize in. You figure out the target that you want to hit, then you bend the data to make it fit your model. If you are clever enough at statistics, you can cause the data to bend simply by changing the rules and the assumptions.

these estimates and determining how close (or off) you were in the original cost accounting? Even if you don't check every single invoice, you certainly need to perform comparisons on at least a random sampling of recent and current work. There's a solid argument that suggests, once you've done the same job a dozen times for the same client, the budget-to-actual-cost ratio will remain relatively stable within the same workflows. Armed with this information, it much easier for estimators to notice potential problem areas that tend to recur. Take, for example, the RIPping process. Look at estimates and costs for jobs from two years ago (if you still have the paperwork) and look at some current projects. The time estimated for output is very often "curve fit" to make the job appear profitable—even though delays and errors at the imagesetter can quickly eat up a lot of time and money.

Financial Considerations

Often, unanticipated problems impact jobs too late in their workflows. Twenty jobs with 10% (actual) profit margin are easily offset by one project that fails to output the first time. Without a solid grasp of profit potential—on a project-by-project basis—overall margins are often less than those forecasted. The focal point of any estimate: the actual cost comparison used to determine realized profit when the invoice is generated. Speak with your accounting personnel to determine how much time it would take to run a simple comparison, assuming the original estimate and any purchase orders were routed to their desk before the preparation of the invoice. It shouldn't take more than a few minutes per invoice to check what the job really cost—and to raise a red flag if profits fall below an acceptable level.

Resource Considerations

The first requirement in setting up a feedback system is a spreadsheet or database that allows easy entry of forecasted and actual figures. Your accounting department is already performing similar functions. Creating such a spreadsheet or report generator is a simple process and shouldn't take an experienced operator more than a few hours to create, verify, and distribute. From there the cost is solely based on the time it takes for a person to enter the data and output a report. These reports should be combined periodically to provide management with vertical data (columns derived from multiple records) telling you how successful you were at maintaining your profits. Armed with this report, estimators can easily isolate areas of concern and discuss them with sales, customer service, and any other departments that are involved in that specific workflow.

Observations

A project output on one imagesetter might prove more profitable than the same project run through a different workflow. The more estimators know about various workflows, the more likely they are to be accurate. The effect expands with time—they just get better and better. Some workflows are more profitable than others, and estimators must know how to quickly identify them. Each department should make a concerted effort to improve communications between estimators and the daily goings-on within their domain. Many managers fail to encourage this type of interaction and pay the price in the end.

Another thing to consider is exactly *when* your estimators learned their trade. In many cases their knowledge set is based on conventional (read: analog) workflows. Although they know what an imagesetter is and how it works, and they surely know that the bulk of the mechanicals are prepared electronically, they may be out of touch with the digital workflows you use now. Preflighting courses can aid in their understanding of these new, and perhaps unfamiliar, processes.

P I P

8

Preflighting Level 1

ABSTRACT

Preflighting is the process of receiving and checking the completeness of incoming electronic mechanicals and attempting to anticipate production problems in a structured and repeatable manner. Level 1 preflighting involves a simple review of incoming client materials to ensure that all items required are accounted for before work is begun. Typically this does not include opening files.

STRATEGIC APPROACH

A "preflight form" should describe the project in great detail. The salesperson is usually responsible for making sure the client has filled out such a form. With the preflight form, a directory listing, and a laser proof of the project, the salesperson and CSR can identify missing elements or information quickly and easily—if they know what to look for. This first stage of preflighting will catch many common errors before they affect workflow on the shop floor. A good preflighting program encourages compliance with accepted workflows by providing feedback to salespeople and clients alike.

FINANCIAL CONSIDERATIONS

Letting an incomplete project move past the creator, past the salesperson, and often past the CSR before realizing that something is missing has a substantial impact on smooth and effective workflows in the factory. When a job stops for any reason, it costs money. This cost can only be determined on a site-by-site basis. To institute a multilevel preflighting strategy, you must first train the creator to use proper page construction and file preparation. You must provide the salesperson and the CSR with training in how to read the preflighting report, how to compare a laser proof to a directory listing, and how to approach this data collection in a standardized and repeatable manner. Professional training organizations normally charge between $400 and $750 a day (per person) for structured training in preflighting. Such training should include the forms your salespeople will need to collect the required information.

Resource Considerations

To institute a sound preflight program, you must impose a good deal of structure on how your clients prepare their files—particularly where it relates to fonts, color models, image sizes, OPI requirements, trims and bleeds, complexity of illustrations, special screening or color requirements, and other issues specific to your client profile. Next, you must train your salespeople and customer service representatives to meet the requirements of a Level 1 preflight. Finally, you must design and supply such file repair and preflighting forms as your situation requires.

There is an increasing effort by developers to provide digital tools that help facilitate preflighting. Among these are applications that check the veracity of the PostScript code generated by the mechanical and others that gather components such as fonts, illustrations, high-res scans, or OPI-referenced separations.

Observations

For purposes of developing a workflow strategy, consider preflighting as a two-level process, and file repair—the correcting of problems identified during a preflight—as a third and discrete function.

Working with structured forms that you supply, a design site should be trained to supply the salesperson from the manufacturer with the following items: the components of the project; a report showing the fonts used; images contained in the mechanical; trim sizes, bleeds, and other special requirements; a laser proof of the project printed at full size and tiled to multiple pages if necessary; and a directory listing of all the components, printed from the disks being sent to your site—not from their local drive.

The salesperson and customer service representatives should be fully familiar with these requirements, and any others specific to your particular business. Every job should contain exactly the same number of components—not page elements, but from the standpoint of hard-copy forms, reports, and proofs. Even without a computer, this first level of preflighting will often catch errors of omission. Missing elements, incorrect disk contents, version errors, and other component-related problems cause a majority of workflow failures.

All of this depends, naturally, on the willingness of the client to comply with what you determine to be sound page-construction practices. Compliance requires motivation. How do you motivate the client? You might consider structuring costs so that files constructed according to your instructions are subject to discounts, while projects that continually cause problems cost more.

You can often trace repetitive problems to specific individuals—ones who refuse to use techniques known to produce smooth workflows. Don't underestimate this factor. Discounts work to gain compliance from the most obstinate clients—especially if you have proof that their projects always take more time and cost more to process. Going to a designer's manager and showing them how much it costs their organization to accommodate certain personnel will get results—but always be professional, no matter how difficult it might be.

PIP 9

Preflighting Level 2

Abstract

A Level 2 preflighter opens files and analyzes each component in a structured manner. Page layouts, illustrations, and image files are combed carefully to reveal specific problems in need of repair. The Level 2 preflighter must be software savvy and thorough.

What follows are two examples of Level 2 preflight forms that you can modify for your situation.

Preflight checklist

PreFlight Job Checklist

Job# 82999 Petstore 10/9 Sale | Date PreFlighted 11/14/93 | PreFlight Operator TB
Image Size 10 1/2 x 10 | Spread Img Size 10 1/2 x 20 3/8 | Server F2V5
Page Size 11 x 10 1/2 | Spread Pg Size 11 x 21 | Checklist Page 1

File	Image Size	Page Size	Color	Art	Lasers	....	....	Notes
~~Pg08&01.Base~~	~~☑~~	~~☑~~	~~☑~~	~~☑~~	~~☐~~	~~☐~~	~~☐~~	DELETE THIS PAGE
Pg02&07.Base	☑	☑	☑	◎	⊟	☐	☐	"PurinaLogo.eps" missing
Pg06&03.Base	☑	☑	☑	☑	⊟	☐	☐	
Pg04&05.Equine	◎	☑	☑	☑	⊟	☐	☐	Spread Img Size ~~10 1/2 x 20 1/2~~
Pg04&05.No.Eq	☑	☑	☑	☑	⊟	☐	☐	
	☐	☐	☐	☐	☐	☐	☐	
IN 11/16	☐	☐	☐	☐	☐	☐	☐	
Pg08&01.Base.t2	☑	☑	☑	☑	⊟	☐	☐	
	☐	☐	☐	☐	☐	☐	☐	

Notes

11/14 – Did not run lasers because of time. Told csr (BW) about missing art and bad images. She said ignore missing art and fix image size problem by decreasing the gutter.

11/15 – Delete page 08&01–New one coming 11/16/93

11/16 – Received new Pg08&01. Gave a .t2 suffix.

Job Checklist 1.0 09/07/93

Trouble form

Please print...

This document must be carefully filled out whenever you have printing problems.

☐ Chooser set to: ______________ Job No. ______________

☐ LaserWriter Driver used: ______________ Client ______________

☐ Printed from application: ______________ Job Name ______________

Attempted:	Error (list exact error; see suggested solutions below):
☐ Print all pages/all colors	
☐ Print one page/all colors	
☐ Print one page/one color	
☐ Print direct	
☐ Restored RIP	
☐ Save As to make new file/Print	
☐ Made sure: Split complex paths of	
☐ Made sure: No Larger Print Area going to hi-res	
☐ Made sure: No "Printer Options" checked	
☐ If many fonts, tried Unlimited Downloadable Fonts	
☐ Simplified clipping path of	
☐ Changed flatness of	
☐ TIFF converted to EPS	
☐ Type converted to paths for	
☐ Recolored in original program	
☐ Changed OPI .lay file	
☐	
☐	
☐	

GENERAL REPORT: Time spent resolving printing conflict was about _____ mins. It is ☐ billable ☐ not billable

Overall analysis of problem(use back if you need more room):

Proper solution was:

PostScript errors	**Suggested cause/solution**
Limitcheck	
fill, eofill, flattenpath, stroke	Autotraced graphics; increase flatness value (up to 35 for 2400 output)
image, color impage	Too complex clipping path; simplify!
moveto, lineto, curveto	Too complex paths; manually split paths. Increase target resolution.
Offending Command	
$%# or nothing*	Corrupted Tiff; resave file and replace in document.
string, array, dict	VMerror; restore RIP. Send file w/ unlimited downloadable fonts.
nothing	Font problems!
Dictstackoverflow	Nested fonts or eps into eps. Convert type to paths; recompose.
Gsave	Delete unused colors, convert type to paths, reduce nesting.

Strategic Approach

There are limits to the possible combinations of people you might assign to the preflighting and file repair functions. Many sites combine the two levels of preflighting with file repair and output, making the entire process of analyzing, fixing, and RIPping the file the responsibility of a few select individuals. Combining the first level of preflight (which checks completeness) with a second level (opening, analyzing, and identifying component-specific problems) imposes more structure on each process and offers the potential for reducing the number of errors that occur past this point.

Financial Considerations

Besides the administrative costs of creating and producing the necessary forms and the expense of training salespeople and CSRs in Level 1 preflighting, you probably already pay for the most expensive item in this approach to multilevel preflighting—the technical expert. Level 1 preflighting is a "pre" preflight—it identifies and often solves simple logistical problems that shouldn't be the responsibility of your most expensive technical people. The second level of preflighting—the one that requires an intimate working knowledge of how programs work—is left to the experts.

Resource Considerations

Many managers make the mistake of dictating, as opposed to participating in, changes. Managers need to get involved. They need to get techies to buy into the idea of moving some of their responsibilities upstream. Although some may resent anyone else doing what they feel they're uniquely qualified to do, many will be relieved at the prospect of not encountering missing fonts, lost or incorrect images, and incorrect trim and bleed specifications. If all they have to do is check colors, traps, and image components, they will have more time to do a better job.

Observations

Having gurus around your shop is important. However, despite the fact that most producer sites employ them, there are some problem-solving tasks they needn't be involved in. That is, the easy stuff. Gurus sometimes "overtech" a problem in search of a solution. This can result in bottlenecks during preflighting, and many problems still escape the guru's notice and result in unacceptable output. By spreading the responsibility for preflighting and incoming file analysis over several different individuals, with different levels of expertise, you limit your exposure to having too much technical wisdom in too few hands.

Don't assume that you need to have a ten-year desktop veteran to effectively perform the more technical Level 2 preflighting. Most work arrives at the factory with Illustrator or FreeHand artwork, Photoshop images, and Quark Xpress or PageMaker mechanicals. (Although there are certainly projects outside this construction model, they are few, and usually represent unusual situations.) If the majority of your work fits this profile, then training relatively inexperienced individuals to do Level 2 preflighting is certainly possible.

PIP

10

File Repair

ABSTRACT

File repair is the process whereby a highly trained and experienced individual, working from information prepared by the preflighter, fixes problems in the file that might stop or delay its output.

STRATEGIC APPROACH

Perfect preflighting should catch every imaginable error, every time. We say "imaginable" because something new always seems to crop up to ruin output. Upgrades and system changes are examples. If new technology introduces a workflow glitch, preflighting will identify its frequency, letting you adjust your file repair techniques to address it in a repeatable manner.

Considering, though, that a vast majority of problems are

ones that you've seen before (fonts, image linking, page geometry, etc.), you should worry about the common problems and leave the esoteric ones for last. Remember, you can only base standards on commonalities—not aberrations.

Having had considerable experience in focus groups, we know that many techies often incorrectly identify priorities when it comes to addressing workflow problems. They often seek technical solutions (new equipment, software, etc.) when process control is what's really needed.

Financial Considerations

Structured preflighting is greatly enhanced by cross-training staff members. Accurate preflighting directly affects how soon errors are caught. The longer it takes you to catch an error, the more it costs to fix. Output that can't be used is obviously the result of nonproductive time. We know of a case where, upon the termination of an output manager, the owners found reams of bad film stuffed into the space between the roof and the drop-ceiling panels. Although we hardly think this is happening at your plant, it drives home the financial impact of letting jobs slip all the way to the output event before someone realizes something isn't right.

Resource Considerations

Staff machines must be equipped to open any component that might be found in the mechanical, including high-resolution images if they're part of your workflow. This will become more important as designers move toward working with their own high-res images. You should avoid situations where only the scanning and imaging stations are able to open client files. Otherwise, a preflighter can't check vital information such as color models, apparent artifacts or casts, and other problems that crop up when designers start doing their own retouching. Add to the cost of equipment, a department needs to maintain the high cost of experts needed to do the work. Fortunately, most sites already employ one or more gurus who are often self-taught. On the other hand, if you're able to clearly define a set of preflighting skills, you can teach them to someone.

OBSERVATIONS

Consider the FBI approach to forensic evidence analysis: have a room full of equally skilled and trained individuals perform either Level 2 preflighting or file repair. A file repair task would naturally check the results of the Level 2 preflighting. All file repairs would be double-checked by another person in the department prior to output. This way, projects move from one desk to another and rarely result in unacceptable output.

File repair is a tedious process, and file repair operators are prone to burnout and boredom. Be on the lookout for signs and give people a break from the routine—even rotating them into another job.

PIP 11

Catching Output Errors

ABSTRACT

Output occurs when a file is actually sent to an imagesetter or other output device. It's critical that every aspect of preflight analysis and repair have been completed before outputting begins.

STRATEGIC APPROACH

Have you ever tracked the number of jobs you output that aren't acceptable? At many sites—yours, perhaps—the percentage of "first-time-accepted" film is lower than it should be. In some cases, it's abysmal. What is acceptable? Many managers would be happy with 95%. We feel it could be 98% if upstream processes are in place.

The reason pages fail to output properly has to do with quality control. Look at workflows today. All too often, too many changes are made to files before they arrive at the output department, with little or no QC. In most cases quality control occurs when the film is pulled, which usually results in rework. Is this the case in your workflows? While there are many paths a project might take, making time to check and ensure the accuracy of prior processes is critical.

Other areas to pay strict attention to are wastage and spoilage. Wastage occurs when film is exposed, but later thrown away. A postage stamp imaged in the middle of an 8-x-10-inch sheet is a drastic example of wastage. Think of the thousands of inches of material you trim and subsequently discard. Why not just throw money in a shredder? Try grouping or ganging jobs to maximize film usage. It can be done manually (using a page layout or illustration program) or more automatically, with a one of the utilities available from manufacturers of imagesetter and output devices.

Spoilage refers to scratches, dents, and dings that render perfectly imaged film unusable. Environmental conditions and improper handling are usually the culprits here, and can only be controlled by behavior modification—in other words, people must keep the working environment spotless and be very careful handling output.

Financial Considerations

"I can't pull a laser proof. There's no time. I have 40 jobs on my desk and they're all late." If you've ever heard this argument from one of your employees, chances are you do too much rework. The cost of instituting better process-level quality control is measurable in many areas, but let's start with film wastage. If you throw away 15% of the film you buy, it doesn't take Miles Southworth to figure out you need more and better QC earlier in your workflow. Figuring out how to reduce wastage is worth almost any price you pay. In the end, you will gain far more than you spend.

Resource Considerations

Again, solid preflighting is the key to proper quality control before a project is output. It requires trained staff and the equipment they need to properly check for errors along the workflow stream.

OBSERVATIONS

Keep in mind that when you pull film that isn't acceptable, it costs far more than the money you've lost on materials. There is a popular theory within the automobile industry that gives us some insight into the bad-film syndrome. It's called the Price of Nonconformance. Simply stated, it says that if you build a car and the transmission costs $1,000 to build, shipping that car with a faulty transmission will cost ten times the original amount to recall and repair. The next faulty iteration costs ten times as much as the last fix—and so on.

Price of Nonconformance Phillip Crosby, in his book *Quality Without Tears* (McGraw-Hill, 1984), estimates that the cost of failure in a manufacturing business represents 20% of gross sales; in a service organization, it can reach 35%. Our industry falls somewhere in between manufacturing and service, with very few firms having profits as high as 30%.

If this theory is even fractionally true for other manufacturing environments (and there's no reason to believe it isn't), then a page that should cost $20 in materials and time to do right, costs close to $200 to do wrong. Do it wrong again and the loss skyrockets to $2,000. Sound too high? If you have ten jobs to do in a shift, and only seven output properly, the next shift's initial capacity drops to seven—and those seven are likely to contain their own percentage of faulty pages. Keep calculating for a week or two and you'll see the money oozing out. That green stuff on the floor is your lost profits.

PIP 12

Organizational Structure

ABSTRACT

Many organizations are top-heavy, with too many people spending too much time making management decisions. In such cases, middle management seems to hinder workflow rather than facilitate the smooth movement of work from the front door to the shipping dock. Remember the definition of productivity: time spent actually generating income. Management is—by its very nature—nonproductive. Some is necessary, but not nearly as much as is commonly in place.

Strategic Approach

What is middle management? How does your organization function? A positive step in addressing the balance of responsibilities in your shop would be to draw a realistic organizational chart. First, determine reporting structures at the lowest levels—do all the CSRs report to one manager? How about strippers, system operators, and salespeople? Are there technical people in your organization who really don't report to anyone, since hardly anyone knows how to measure or critique their performance?

Balancing responsibilities should be the real target of any organizational change. If you agree with our definition of workflow, then the individual responsibilities (for specific processes) are clearly the domain of the worker—not the manager. If there's enough structure in the workflow processes (they're well defined, mapped, and taught to everyone), then the need for management on a departmental basis diminishes. Some might say it shrinks when those well-trained, disciplined individuals become part of a multidepartmental team. Place responsibility for decisions in the hands of the people most affected by those decisions.

Financial Considerations

Managers cost money. Since they're often drawn from the most talented employees (within a specific discipline), their movement from staff to management usually causes a direct reduction in productivity. A good example is the highly successful salesperson who fails miserably at sales management. Your sales go down (because they're filling out paperwork and having meetings instead of selling) and their income suffers. How much do you currently pay for middle management, and would those people be better off actually participating in the manufacturing process (as opposed to the paperwork and meeting thing)?

Resource Considerations

Many managers are called managers for reasons having little to do with their leadership abilities or their actual contribution to workflow. If you choose to reduce the number of middle managers in your organization, some will not make what they perceive to be a downward move in their careers to become a part of the staff over which they once held sway. They will not make the transition and you will probably lose them to other companies—competitors who believe that lots of middle management is a good thing. Hiring new people is one option. Attrition is another.

Many companies actually inherited their organizational chart. "Sam's been here for 200 years. He ain't gonna go for that new imagesetter—I can tell ya. I know the guy. He has his heart set on that new set of Civil War pistols for the lobby, and won't be happy till he has 'em." If you want to improve profit margins, you may have to use tough love.

Observations

Are we sounding like French revolutionaries out for the heads of every manager in our business? We really don't mean to, and obviously there are many managers doing a wonderful job. However, reducing the number of managers and spreading responsibilities for specific processes more horizontally onto the shoulders of the people who actually do the work seems to have worked for the sites that have tried it.

The size of an organization naturally has something to do with the latitude you have, as do the process controls that are in place throughout the workflow. You can't simply fire every manager or assign them a scanner to run and hope that everything will work out fine. Not all management is bad, and not all overhead is wasted. However, if middle managers insulate upper management from the floor, then you must either change the structure or continue to pay for keeping someone's personal sandbox brimming with toys.

PIP

13

Cross-Training

Abstract

Cross-training refers to developing staff capable of performing each others' jobs. Some people take very well to this concept because it puts them in touch with the totality of the business. From the standpoint of workflow, it dramatically improves your agility and improves the quality of feedback you receive from individuals in the organization.

Strategic Approach

In Chapter 4, "Mapping Workflows," we invented the perfect employee—one that knew how to accurately execute every step in the process that he or she was responsible for. Cross-training is aimed at producing just such an employee. The most successful approaches to cross-training that we've seen are based on an incentive/cost strategy. Employees who want to learn about other aspects of the business are encouraged to do so, and offered some sort of incentive compensation if they successfully complete the necessary training (and, if necessary, apprenticeship). We also think that at least some of the time devoted to this training should come from private time. This ensures that the employee identifies with the cost you're willing to incur to teach them what they need—or want—to know.

Financial Considerations

Cross-training isn't something that pays back in hard dollars immediately—like preflight training might. The cost for turning a stripper, let's say, into a fluent Photoshop retoucher is probably the same as it would be to turn anyone else into a good retoucher. The difference here is that the time the stripper spends to learn a new skill can't come solely out of your routine schedule—or you would have to hire another stripper to take their place in the interim. At the very least, the cost in time should be shared by the student. Once they complete the training, they're clearly worth more—both as skilled individuals capable of doing more than they could before and, even more important, as members of a cross-discipline team. If individuals are personally aware of workflow profitability, they are more likely to make solid decisions about which training to take.

Resource Considerations

In general, you have to get people to buy into an idea—no matter how much it ultimately benefits them. To approach cross-training, you have to have secure employees who recognize that the more they know, the more they're worth to the organization. And management needs to set up a reward system to encourage cross-training.

Observations

Attrition is likely to occur more frequently in our business than ever before. Strippers and camera operators are heading down the same road that was recently traveled by typesetters. Digital imposition and assembly is increasing and the trend is only likely to continue.

Facing the loss of their craft, many people in the color industry are justifiably worried about what they're going to be doing five or ten years from now. Cross-training provides a way to apply the knowledge they have to newer processes and techniques made possible by new technology. They're usually willing to spend the time and often represent your most loyal and longstanding employees. If you make sure not to take the entire "hit" in terms of time, retraining and cross-training (two sides of the same coin) can substantially benefit your organization.

One option for cross-training is something we call "shiftsplitting." An employee works her shift—or a portion of it—and then moves to a different position for additional hours. Several sites we know have used this strategy quite successfully. The person isn't paid overtime—just the standard (or slightly reduced) rate for the cross-discipline position. It can be thought of as an after-hours apprenticeship program, where the employee is paid to learn.

PIP

14

Job Descriptions

ABSTRACT

Job descriptions should describe job requirements. They define the employee's responsibilities as they relate to your various workflows. This has broad implications—in hiring, training, cross-training, workflow development and design, compensation, and employee satisfaction.

STRATEGIC APPROACH

Do you have job descriptions for all the employees within your organization? If not, this is the time to write them. Start by assigning a title to each person in the organization. Don't leave yourself out. Once you have a title assigned to every person, write a short paragraph about each job, providing a brief description of what the company expects that person to do. (In many cases, stark contrasts between reality and perception emerge.) At this point, you have the basis for a redefinition of job descriptions, required skills, and what you should look for in your search for new hires.

FINANCIAL CONSIDERATIONS

Certain job descriptions—especially those rooted in conventional manufacturing techniques—don't have a place in a digital manufacturing environment. As we move from a craft-based to a process-oriented manufacturing environment, some jobs will vanish. Typesetters, once so important in the graphic arts workflow, are a perfect example. There are others as well: engravers, dot etchers, camera operators, and strippers (currently the most highly endangered species). Once job descriptions are in order, potential losses due to attrition and areas for retraining will become apparent.

When you've completed your analysis, it will aid you in your efforts to train (or replace) current staff whose job descriptions or skill sets don't match what they're expected to do.

Resource Considerations

If you have a human resources (HR) department, it should write job descriptions and responsibilities. In a mid-sized or smaller organization, management should take on this task. Even in situations where a formal HR department does exist, it's still vital that management stay involved.

Do you currently support positions that can best be described as vague? What's a Mac operator, for example, or a guru? A solid approach to job definition should not only eliminate such ill-defined positions, but should also take into consideration your staff's personal growth: where can they move? What are their career paths in your organization? If there are none, consider strategies to maintain a consistent workflow while dealing with continual turnover.

Observations

The industry clearly needs a set of well-defined skills standards. The Graphic Arts Technical Foundation (GATF) recently developed a set of skills standards that can serve as a starting point to help you better define and anticipate required skills.

It is, however, only a starting point. Any manager in our business with a working knowledge of his own specific workflows, is challenged to define job descriptions much more accurately than is currently the case.

Take a minimalist approach. Develop a mental picture of a functional team able to start and complete a project within a profitable workflow. Then see if you have people with wrong titles on their business cards or who occupy desks in the wrong part of the plant. This exercise might surprise you.

PIP

15

Management Involvement

ABSTRACT

Managers should always be in touch with the organization they're managing. The key role of management is to provide clear objectives that are understood by those who need to fulfill them. Having unclear, unrealistic expectations; keeping personnel in the dark; yelling at the top of your lungs; or saying one thing and doing another fall short as management strategies.

STRATEGIC APPROACH

First, recognize that at least 85% of all workflow problems are management problems. We don't set out to hire bad employees or buy the wrong equipment.

Reengineering workflow requires changing not only processes, but also behavior, throughout your organization. (And beware: changing someone's behavior is a lot harder than changing the way they do things.) Outward behavior is only the tip of the iceberg—the portion underwater that anchors the iceberg is greatly influenced by cultural background, social attitudes, and a person's level of contentment.

If you want to be a leader, you have to lead. Management very rarely sits at a workstation, or maneuvers a forklift at the loading dock, or joins the CSRs for lunch. But why not? A bit of camaraderie on your part goes a long way in opening and building lines of communication with your staff.

FINANCIAL CONSIDERATIONS

Do you currently track, on an hourly or weekly basis, how much individual executives cost your organization? Based on a conservative estimate, a large shop typically spends $500 or more, per person, per day, to stock the shelves with vice presidents, presidents, department heads, and other high-flying positions. Do all your executives provide the return on investment you are paying them for? If not, you may have one too many managers. At $40 or $50 per hour, an executive should do anything (within good taste, that is) necessary to make the company profitable.

Resource Considerations

Beware of the "I don't do windows" attitude—it's often rife in our industry. Designers don't do typesetting, print buyers don't do preflighting, strippers don't do scans, and executives have better things to do than "do" a day as a CSR, right? Wrong.

Observations

Try "walking a mile in someone else's shoes" before making any strategic decisions about how to run your business. You might be among the thousands of owners and managers that have successfully worked their way up from the factory floor to the executive suite—but that was a long time ago. Make sure your involvement in developing a solid workflow strategy doesn't stop at assigning the task to someone else, or simply leaving this book on their desks (although in the opinions of your humble authors, you should probably buy each employee a personal copy).

According to generally accepted strategic planning theory, the single most important cause of reengineering failure is lack of management involvement. Analyze a week of your own time. Keep a diary and review it over the weekend to see exactly what you do in a typical week. A good percentage of your time was probably spent in meetings with middle managers and outside vendors (bankers, lawyers, accountants, equipment salespeople, film salespeople, and so on). Both activities are questionable in terms of spending time productively.

PIP 16

Determining Processes at the Client Site

Abstract

Quality control must become a subprocess at every stage of the workflow—starting when the job is first created. Reserving quality control until after film is output means you'll catch mistakes, certainly, but if you had caught them earlier, you wouldn't have to do as much time-consuming and expensive rework.

Strategic Approach

At most sites a job comes in, it is staged, preflighted, repaired, combined with high-resolution images, and output. To minimize mistakes and problems, quality control must begin before staging; in fact, while the job is still in the hands of the client. Consider having production teams—in their entirety—visit key client sites. Tours can be part of your preemptive strategy to reduce errors as early as possible. Select a few clients whose projects normally run through the shop with little difficulty, and a few others whose projects are routinely problematic. Have your salespeople initiate a tour and help choose a contact at each of the client sites to act as a "tour guide" for the group.

Financial Considerations

A tour is generally disruptive to the client and the visiting team. It's a pain in the neck to organize and execute; but the potential return is dramatic. A team visit costs the creator and vendor the time it takes out of everyone's day. Since a visit is a courtesy granted by your clients, first discuss the benefits they will realize directly, to offset any disruption it may cause.

Resource Considerations

What happens to the projects that are on people's desks the day they drive across town in the company van? A well-disciplined and cooperative organization can spread out the day's work to cover the traveling staff and ensure that everything will continue to move through the workflow.

Simply visiting the site isn't enough by itself. While the issues uncovered during the trip are still fresh in the minds of the team, take the time for a thorough debriefing. In addition, assign someone the task of writing a follow-up report for general distribution so that as many folks as possible can benefit from the experience.

Observations

The Video Alternative
We've recently seen shops that have made good use of videoconferencing in lieu of group visits. While not the same as a site visit, a videoconference can establish the human, face-to-face relationships you ought to encourage. There's something warm and fuzzy about bringing in the scanner operator or the press operator to discuss specific problems or procedures while the CSR and the prepress manager are standing by.

Group visits help you to better understand your clients' workflows. It also helps reinforce the ideas of a workflow map specific to the creator site and that they too have profitable and nonprofitable workflows. Manufacturers are more likely to have a formal process map than an agency or publisher. If one doesn't exist at the client site, offer to help the client develop one. Simply analyze those workflows as you would analyze your own.

Jobs start with designers. Do they attempt to do too much? Do they trap their own illustrations or convert RGB files to CMYK during the creative process? If so, it's likely that errors are introduced during the design phase—errors that you are now paying to fix when the job arrives. Do designers and client-based production artists understand your needs for specific construction practices? Do they understand that you have workflows that produce satisfactory results, and others that don't? Educating your clients and enlisting their support can streamline the workflow process and ultimately saves time and money all around.

PIP

17

Feedback and Reward Systems

Abstract

How does your staff know if they're doing a good job? Are they rewarded for exceptional performance? Is salary their only motivation? If so, you might find that rewards and feedback have far more to do with productivity and profitability than you think.

Strategic Approach

There is a fundamental difference between a raise and a performance reward, both from the standpoint of how much it costs the organization and how it is perceived by the employee. Performance-based reward systems have proven very effective in certain environments, and might very well have a profound effect on yours. Salespeople working on commission know the routine—if they don't perform, they don't get paid. Although a pure commission arrangement probably won't work for the bulk of your staff, reduced salaries combined with increased bonus plans often will.

Financial Considerations

If you go out to the shop floor and announce that you're going to start a bonus program, the results will be very positive—but also short-lived. To attach a cost to your bonus incentive plan, you must first determine what your current profitability is. Then determine what increases will generate money for said bonuses. Decide to whom the bonuses will be delivered, and in what form. Keep in mind, money isn't the only bonus you have at your disposal—additional paid vacation days are often worth more to employees and actually costs you less. Another great incentive is training in areas of interest to the employee (but not necessarily connected to their specific range of duties), such as an artistically inclined CSR who is dying to mess with Photoshop.

Resource Considerations

Bonus programs based on measurable improvements in workflow require that you first determine what improvements you want to happen. Do you want to decrease the time a project spends in customer service? Figure out how long an average job spends in CS and what percentage of those jobs are perfect when they leave. Tell the CS department that if the average time is reduced by 10% and if the perfection ratio improves at the same time, you will send them all to Wally World for the day. Yes, you will shut down the entire department for a day—workflow be damned! See what happens. At a cost of $100 a head for a five-person department (and the cost of losing them all for a day, let's say another $100 per person), you could save 48 minutes a day (10% of the 480 total minutes in eight hours). Multiply those 48 minutes times 20 days and you've added two work days to the month. Are two additional days a month worth a $1,000 day trip?

There are many other incentive possibilities. If your people work in dreary cubicles, getting cool new furniture is a welcome reward. So too is hardware for their systems—new monitors, more RAM, even game software for their home systems are possibilities. How about offering color output as a bonus, or even separations and film? Most individuals have outside interests (we hope) and sometimes have a need for graphics assistance for their church, a club, or some other association. Sure, many employees could easily sneak in at night and secretly work on such extracurricular activities, but doing so with your blessing is beneficial all around.

Observations

We've had the experience on more than one occasion where a client site was having cash-flow problems and faced the prospect of laying off a lot of staff. As an alternative, we suggested that they reduce salaries across the board (starting with themselves and saying so) and implement a productivity/profit-sharing reward structure. In every case the companies were able to turn around and move back into the black—often in much less time than we thought they would. A little analysis will show you why.

When faced with the chance of losing 100% of their income, the staff—as a whole—agreed to accept less money. They knew there were problems—employees often sense something is wrong before management admits it to themselves. Getting employees to accept a reduction in pay, while painful, wasn't as hard as the owners thought it would be. Tying the salary reduction to a potential long-term increase in income was really the trick though, by agreeing to provide productivity bonuses to individuals and departments, should the strategy work to turn the company around, management promised to pay everyone more in the long run.

And last, but certainly not least, remember that a kind word of encouragement and recognition—when it's deserved—is perhaps the most effective reward you can give your workers. If someone does something that deserves praise, give it to them. Being hard-nosed isn't nearly as effective as being kind and supportive. By the same token, you shouldn't cheapen praise by giving it away too freely. Simply be generous appropriately when someone makes you money.

PIP 18

Finding and Correcting Errors

Abstract

Despite your best efforts to improve workflow, errors will still crop up. It's important to establish a mechanism to capture, record, report, and process such errors in a consistent and controllable manner. This is the essence of quality control—use errors as guideposts for future process strategies.

Strategic Approach

A structured approach to finding and correcting errors obviously draws from a well-developed process map. Any event presents your staff with an opportunity to make an incorrect decision. Operating under that assumption, any event on the map represents a potential error. Working from the map, you might try to predefine what errors might occur at any given location. For example, when a job is preflighted, failing to determine if the fonts are included with the job imposes a burden on individuals who are in subsequent positions on the map. Either the fonts got checked or they didn't.

This isn't a blame game we're talking about here. We set out to hire good employees—not careless ones. Responsibility comes with a price tag. Employees often (justifiably) feel that they know what needs to be done better than managers hiding behind closed doors. They're usually right. On the other hand, there are people who love to moan about how powerless they are, until you grant them power. Then they moan about responsibility. In the long run a method designed to catch mistakes will clearly identify these individuals. When you're sure that a certain person (and not a process—or yourself) is a continual source of errors, weed him or her out.

Financial Considerations

Catching errors reduces rework—period. It's not easy to get everyone culturally adapted to accepting the onus for their errors, and attempting to do so can meet with resistance if not handled gently. Some sites use a form that has a space called "other employee" to indicate the source of a problem. Such a form allows everyone a graceful escape from assuming responsibility for a blatant error.

Resource Considerations

A good way to start catching errors is to create an error-tracking form. Don't make the form too complicated. People hate them and they often serve no useful purpose, collecting (as they often do) useless information. Track only what you need to track (we discuss this at length in Chapter 6, "The Database").

A simple error form provides each person that handles the job a place to indicate what was wrong with a project when they received it. By indicating the nature of problems, it serves as a basis for simple statistical analysis of common problems. Are traps always created incorrectly from a particular client? Does a specific preflighter consistently forget to check color values in Quark XPress pages? Does a given scanner consistently introduce artifacts into digital images? Is a specific imagesetter scratching film? It's a fundamental truth that system-related problems can be repeated—if a particular command results in bad output, it will result in bad output every single time you duplicate the process. There aren't too many unknown bugs in Quark XPress or Adobe Photoshop. If errors are being introduced into the workflow, it's either from someone missing something, or because a particular function is applied incorrectly or out of order. A good error-tracking form will help your staff find, identify, and eliminate repetitive errors.

Observations

Let's say a CSR's job is to check the project for completeness and route it to a Level 2 preflighter. That person will look at the laser proof and begin a structured analysis of color objects, choke and spread settings, and the construction of any line art elements, and will carefully analyze continuous-tone elements. A report is filled out and the file is sent for repair and output. The next person in line should expect certain things to have been done by the previous operator, and so on through the completion of the project. If someone doesn't follow through on their portion of the work, it needs to be noted. (See "Ownership of the Project," page 142.)

PIP

19

Time Scheduling and Load Balancing

ABSTRACT

All the workflow planning and process adjustments in the world won't help if there's too much work in the shop to handle with the available resources. By the same token, the smoothest management program on the planet will not make employees cost effective if they don't have work to do.

STRATEGIC APPROACH

Most shops operate under the assumption that employees should work in shifts of eight hours each. If the workload exceeds the capacity of one shift, you simply add another—or rely on expensive overtime to fill the gaps. If this approach leaves too many employees standing around when work is slow, or too few people available during peak loads, than you might consider altering your approach to scheduling.

There are two issues to consider here. The first is the time it takes to process specific functions on computer workstations. This is one area where raw throughput has a dramatic effect on workflow planning. The longer it takes a RIP to chew through a typical project, the longer it takes the job to move from file repair to proofing. RIPping files is rarely strictly scheduled, often happening whenever a project appears ready to output. If your imagesetters are frequently idle, with bursts of major activity (which can't help but create bottlenecks), consider scheduling RIPping times using information from sales and customer service personnel or from project team leaders. If your output people don't know what they have to RIP tomorrow—or even the day after that—then chances are, time is being wasted.

Secondly, what law says that five eight-hour days constitute a shift? Have you considered having teams or departments work four ten-hour days? Or any hours that they feel are appropriate to produce a predetermined set of projects? If you need to move 75 pages through your shop each shift, what do you care which hours between Monday and Sunday the necessary staff decides to work? If a team chooses to stay with a set of projects until they are done, they might decide to work 14-hour days—if that's what the project takes. Rigid scheduling can make people unavailable when they are most needed.

Resource Considerations

Scheduling RIP times is a matter of producing a workflow that encourages more accurate time estimates. As is the case in cost accounting, improving your control over time accounting requires that time requirements be forecasted early in the job and revisited once the project's been completed. Comparing before-and-after time estimates will allow you to develop a sense for where such estimates are accurate—and where they're not. This exercise often points out the equipment you already have that is idle—and if it is, you hardly need to purchase any more. A slow imagesetter working all the time is much more effective than a really fast imagesetter attempting to do a day's work in two hours.

The issue of staff schedules is a bit more complicated to implement. In our opinion, you really need a team structure to have it work. Since a team includes people from different departments with a well-rounded set of skills, they're more able to organize their project load and approach it from start to finish than a manager might be.

Observations

One of the most effective ways to reduce latency—the time projects spend just sitting in a bag or on a hard drive—is to assign a single team of people to start and finish one job.

We often approach scheduling within artificial frameworks. For example, if you have a project on your desk that isn't due until next Tuesday, why work hard to get it done in the next 24 hours? After all, there's other work in the shop that needs to be done today, right?

First of all, try to imagine if every job that came in went right out the next day. Take a look at the case study on Graphics Express (in Chapter 3, "Real-World Operations and Their Workflows"). They process *hundreds* of jobs every night between the late afternoon and the next morning. How? By reducing the latency factor to almost zero. Jobs don't sit around, even if they could. They come in and get completed immediately.

Some would argue that if you give the client a job overnight, they'll expect the same response from that point forward. This has a ring of truth to it—people are, after all, people, and they do tend to want the impossible. If you're concerned about this, yet still want to try alternative project scheduling, just deliver jobs as scheduled, not a moment sooner. If the job wouldn't normally get done until next week, and you get it done tonight, let it sit in the bag completed until the promised delivery time. The latency factor will still affect your cash flow (you can't bill them till you deliver the work) but it will stop having an effect on your workflow. There are worse problems than having all the jobs you need to deliver next week in shipping, ready to go.

PIP 20

Ownership of the Project

ABSTRACT

Do your employees sign their work? Do you sign every project that gets delivered to your clients? Does the thought of doing that make you just a little uneasy? If it does, just think of what your employees would say if you told them tomorrow that they all have to sign their work before they pass it on to someone else in the organization.

Strategic Approach

This PIP relates directly to finding and correcting errors. If we assume that all of your employees have a well-defined set of tasks they're expected to perform, then signing their work before they pass it on shouldn't make them uncomfortable. If you attempt to institute a method whereby someone has to assume responsibility for what has been done and that person doesn't feel good about it, then he probably doesn't feel comfortable with his job description. Although we don't want to point any fingers, this is usually a management—not an attitude—problem. It's easy to think it's an attitude problem, but that's at the core of what we're trying to point out—placing blame where it belongs and recognizing excellence when it happens. People hate to take blame for something they missed or did incorrectly. Why? Because it makes them feel bad and wish it was someone else's fault. This is understandable—but not acceptable if you really want to gain control over the work processes in your plant.

Financial Considerations

The cost of implementing an ownership program in your shop might be the loss of certain employees—some people can't stand signing their stuff. Initials, by the way, are different than signatures. Signatures carry power—initials don't. It's OK to initial the insurance portion of a car rental agreement, but on the bottom line they want the whole thing, right? The same thing applies to this tracking form. Have a signature line—not an initial box. Trust us, there is a difference in perception.

Resource Considerations

Signing errors is one strategy that can't really be tried out in one department before it's adapted to the entire company. Signing (and owning) work must be based on the entire process map, with no holes. If people in the loop aren't required to sign off on what they did, then others who do will assume all the problems occurred where the operators remain anonymous. The signers will be able to (justifiably) argue that they didn't know who did what to the job before they got it—so how could they be expected to get their portion of the job right? If they know who had the job last, they can go directly to that person and ask questions—rather than going to a middle manager who could wait until the next department meeting to address the issue.

OBSERVATIONS

People have a hard time accepting the burden of responsibility if there isn't a corresponding reward system. Very few managers who succeed in using workflow strategies to improve their bottom line do so without a structured approach to problem ownership. There are people, unfortunately, that simply aren't very good employees; they constantly repeat the same types of errors. Signing their work will not change them; it will simply get everyone down line to pick on them even more.

Once people assume ownership for their work, they become sensitive to the needs of their fellow employees—both upstream as well as downstream in the workflow. It will take some time, but eventually a strategic approach to individual responsibility will hone your organization and dramatically improve the first-time acceptance ratio of work coming out of your plant.

PIP 21

Latency

ABSTRACT

Latency refers to periods of time when nothing is happening to a project. Another way of looking at latency in your organization is to determine how many hours a project resides in your factory and compare that figure to how many hours it was actually worked on.

STRATEGIC APPROACH

Latency in most environments is abnormally high. If you can reduce latency, you can increase the potential throughput in your entire organization. The first area you should look at is the "staging" or receipt of incoming mechanicals. In many cases projects come in and, because of some simple omission, fail to be immediately processed. How many people on the team are allowed or encouraged to contact the client to secure missing elements? How quickly is this done? Immediately following the meeting, or two days later when someone finally opens and checks the contents of the disk? Is there a structure in place to determine how long it takes a scan request to move from a project to the scanner? Is the salesperson part of the production team, and therefore able to indicate the number of scans that will be required for a particular job? How long does it take someone to actually issue a print command following the preflighting and repair of a project? How are failed projects integrated back into the workflow, and at what point?

All of these examples represent areas where simple routing and tracking changes have the potential to save time and money. The primary focus of latency studies should be identifying when jobs are simply standing around doing nothing—and having nothing done to them.

Proofing is another source of latency in most shops. Check the time between completion of a proof and the client's sign-off. It's often measured in days. This is understandable because the client's schedule is determined independently of yours. Soft proofing directly targets this problem. If the client agrees to view and approve projects on a monitor, then the window of opportunity to gain the go-ahead for the work is dramatically expanded. Furthermore, remote hard-copy proofing can also expedite workflow. In this case, you supply the client (or they purchase independently) an output device that's capable of producing near-contract proofs. Using high-speed telecommunications, your employees transfer a finished mechanical to the client for output on the proofing system. Once viewed, the client can give you permission to complete the work.

Another way to reduce latency is to examine the granularity of the work. *Granularity,* in simple terms, means breaking a project into more manageable chunks. Have you ever had a 50-page project waiting for one scan before it's output? Why? Depending on the nature of the work, break it into signatures or single forms and output what's ready to be output. Process and proof the work. Delaying 49 pages until you can do the 50th costs you time, money, equipment availability, and human resources.

Financial Considerations

Controlling latency requires measuring latency, and this is something that may best be done by having someone spend a few days looking at files, times, and dates, and actually walking through the process map with a few projects. You might consider using a middle manager from any number of departments—customer service seems a good one. Controlling latency is really a matter of watching, tracking, and identifying major points of latency and then rearranging or grouping similar tasks. If you consider grouping similar tasks from disparate projects, it becomes easier to determine where and when specific groups of tasks should be routed.

Among the more important financial considerations is that of invoice or billing latency. How long is the delay between the completion of a project and the time it's invoiced? Ideally the invoice should ship with the work—never wait too long afterward to send an invoice to the client. Interim invoice generation as a job proceeds is an option that works for many shops. The more you know about your workflows, the more you should be able to accurately forecast when the job will be done and what charges will be incurred. Theoretically, you could write the invoice before you do the work. This might be a little far-fetched, but the concept is sound. Since you usually estimate a price to the client before work begins—the only thing that you might not have a grasp on is what it will actually cost you to produce including AAs.

Resource Considerations

In many cases projects are processed in a single path through the shop. One approach to reducing latency would indicate that jobs might be broken into components and run through a specific workflow process as a group.

A second approach requires the involvement of a team in the earliest staging of each project. This alone often reduces latency because more than one department is able to discuss the completeness (or lack thereof) of every project. You might go so far as to make calls to clients based on Level 1 preflighting during a break in the staging meeting.

Always seek to reduce the number of hands that touch a project. The more a job is handled, the more it sits around. The more it sits around, the more you pay to warehouse the work—even in the short term.

Observations

Latency occurs when something that could be done right now is delayed until later. A print server (discussed later in this chapter) is an example of controlling latency using software. A Chooser-level print server capable of queuing projects into appropriate proofing or output devices controls latency common to human-generated print commands. There is no way for a person pressing the print button to possibly compete with the effectiveness of an automated process. Again, the process controls upstream (in the form of proper planning) are vitally important to controlling latency downstream.

Another area where latency has a major impact on workflow is in the proofing cycle. Some shops combine scanned images automatically into large-scale random proof sheets to shave time from schedules, thus reducing latency. A software routine, or script, is created that watches the folders into which scanned images are saved. When an image appears in the folder, the software measures it. When the program has determined it has enough images at the proper size, to maximize the image area of an available imagesetter, it automatically combines the images, labels them with their job number, and outputs the film. The film is used to generate conventional proofs, which are checked carefully before going to the various clients. Knowing that the scans are acceptable for a given project reduces the need to proof composite pages—reducing latency (the time it takes to make a composite proof and get approval) dramatically.

PIP 22

Trapping

ABSTRACT

The choking and spreading of adjoining elements, or trapping, remains an area of constant concern to the service provider. Trapping is device- and product-specific, and should not be done by the people responsible for putting the pages together. Trapping must be automated and moved much closer to the RIPping event than is currently the case in many environments.

STRATEGIC APPROACH

There is, perhaps, an "ego factor" to consider in addressing the trapping workflow. It suggests that in the early days of desktop, designers had to figure out how to do their own traps because their vendors were less than cooperative in many cases, and were severely limited in what they were able to do to accommodate traps once the page had been constructed on the desktop. These same designers now feel that giving up the responsibility for trapping their own files somehow limits their control over the creative process. They're wrong.

Clearly, time spent trapping (or doing anything that isn't related to the actual creative process) is time not spent thinking about design, editorial, page layout, illustration, photography, or visual balance. As a service provider, you need to move back to a time when designers totally ignored traps—and left the process up to you (or now, more accurately, to your software).

Financial Considerations

There is a cost associated with buying trapping software. Any decent package with the features you need costs upwards of $2,500. Relying on the designers to execute their own traps, or making trapping a common part of your preflight/repair cycle, is far more costly, though. Designers miss traps far more often than does, let's say, TrapWise from Adobe. For that matter, a production person trained in press considerations is less likely to introduce or miss errors in choke or spread requirements. Badly trapped film falls into the "Not Acceptable" category and must be discarded. Not that it is all the time—a stroll through the magazine rack will prove this. Many jobs are simply run with butt traps—perfectly aligned edges with no press-shift compensation at all. If this endangers your relationships with your clients (because they care about stuff like that), the downside cost is considerable.

Resource Considerations

There are several scenarios that produce proper trapping. Each has its own benefits, and naturally you have to choose which method or combination of methods is right for your site. Roughly speaking, they fall into three categories: manual, semi-automatic, and automatic. Manual methods call for a skilled operator, working from well-prepared preflight reports, who creates each trap individually. For page-based elements, the operator often uses the program's built in functions; for illustrations she opens the original files and either rebuilds or repairs the artwork with traps properly assigned. The semiautomatic method has the operator using software such as TrapWise to choke and spread illustrations, while relying on application tools to repair or prepare traps specific to page elements. A third alternative utilizes RIP-based trapping routines, usually built on edge-detection technology, which do their best to guess what you need. Lastly, you can—through structured training—teach clients how to build artwork that doesn't need traps.

OBSERVATIONS

In our observations, we find that many sites still rely either on their clients or on very highly skilled internal staff to manually (on the computer, that is) trap elements one at a time. This is a very job-specific process and is difficult to do in a highly repeatable manner. It simply doesn't work well, and raises rework factors to new highs. People miss stuff, even the skilled ones. The less they know about how a specific press works, the more likely they are to do the trapping incorrectly. The best method, in our experience, seems to be semiautomatic and automatic trapping workflows. They cost a little more to implement but quickly pay you back.

Long ago, the buzzword in our industry was *device independence.* You could create a file on your PC and send it to anyone with a PostScript device. Voila! Great output every time. Unfortunately, printing presses are very specific devices—each one is a little different, and they're affected by environmental conditions to a much greater degree than are digital files. Traps (and many other things) are extremely press specific. You'll need more trap on a high-speed web than on a slower sheetfed press. Even two slow sheetfed machines respond differently to different types of substrate and ink coverage. Designers don't need control over traps unless they also want control over the film, press, ink mixes, and humidity in the plant that day. Move trapping as far downstream as you possibly can.

Do you charge to untrap incorrect files? Consider it. Shops that do, find it easy to discourage the clients from making their own traps. Make it a line item on every invoice where it occurs. It won't take long to get a call from the client asking what the heck that charge is all about.

PIP
23

Proofing

ABSTRACT

There are two categories of proofing on which to focus your energy, time, and money. The first is proofs of concept that show design and the accurate positioning of elements. The second category—and the one that is the most important to the service provider—is the color-critical proof.

STRATEGIC APPROACH

Traditionally, we view proofs as close-to-final representations of the appearance of our projects. In a digital workflow, another word for proof is quality-control point—and they occur throughout the process, from design to finished piece.

In the case of color-critical proofs, a random approach to proofing (see Latency: Observations, page 147) can—in certain situations—save a tremendous amount of time and money.

Linearization
For imagesetters, linearization means if 20% gray is specified, the output imagesetter film will have a 20% gray. In other words, the graph of input values and output values is as close as possible to a straight line, where input and output match (for algebra buffs, where $x = y$). Getting that linear response requires good maintenance and regular calibration.

The primary assumption we have to make is that page-specific objects (Pantone™ colors, screened CMYK values, etc.) should reproduce as specified. This is a matter of proper device calibration and linearization. It's relatively safe to say that a quality provider knows when its output devices are producing 30% cyan when "told" to produce 10% (and they know how to bring such incorrect output back into an acceptable range). The second assumption we make is that color—as in photographic or continuous-tone elements—is the critical issue, and one that cannot be assumed to be perfect every time. If you can develop a method for proofing these critical components individually (or as part of a group), then the composite color page proof doesn't even need to be pulled in many cases.

FINANCIAL CONSIDERATIONS

Random, or loose, proofs were the staple of conventional layout. It saved time and money if the client approved the photographs, with the assumption that the service provider guaranteed that the same color would appear in the composite page. With imagesetters, this is even easier than it was in the past, since you can group lots of color images on one sheet, rather than pulling them one at a time as was done in the past. There is really no hard cost involved—it requires only that images are RIPped together. The savings in time and money might prove substantial.

Resource Considerations

The real trick to the proofing process is to have some way to automate the generation of the composite pages. This can be done with relative ease by anyone with a knowledge of AppleScript, FileMaker, or similar scripting and data management tools. It will cost some money to develop a proofing routine, and time to implement a workflow that accommodates such routing of continuous-tone images, but the cost savings can more than outweigh the trouble and expense.

Observations

An increasingly popular practice is to encourage your clients to accept wrong-reading proofs. It eliminates an entire step in the lamination process. If it's color they're approving, the fact that they have to hold the job up to a mirror to read the copy isn't really an issue in many cases.

You might ask yourself whether or not to move to a digital proofing method. There are many considerations here, not the least of which is that digital proofing systems don't output dots, so they present difficulties in approving critical color. Why? Because color might shift during printing due to dot structures that were invisible in the dotless proof. The best digital proofing systems are the big, expensive ones that duplicate dots—often costing upwards of a quarter-million dollars. But smaller, less expensive systems are continually gaining acceptance and increasing in quality and accuracy. Iris printers and 3M Rainbow proofers are examples.

We have seen client sites that are beginning to rely on much lower-cost devices for the output of interim—and even contract—color proofs. 3M has made a stab at this market with its Rainbow Color Proofer. Some sites are getting very acceptable results with this and other lower-cost alternatives to big digital systems.

As new screening technologies gain importance in the workflow, digital proofing might offer an alternative to conventional film-based proofs, especially for direct-to-plate, direct-to-press, and hi-fi color technologies.

PIP 24

Pickups and Archiving Strategies

Abstract

How much of your work is comprised of images or page elements that were used previously in another job for the same client? If handled well, pickups provide an opportunity to boost profits by reducing new work required. How you address storage and organization of client files is crucial to reaping potential benefits—for you as well as for your client.

Strategic Approach

There are three states in which a project might be stored. The first is on-line, where the project and all of its components are immediately available from the appropriate workstations. They live on mounted volumes. The second state is near-line, which describes projects stored on removable media (opticals, writeable CD-ROMs, etc.); these go on-line as soon as the media is mounted. The third state in which a project might live is off-line, which requires access through a dedicated software application. Most tape backup strategies fall into this third category.

Financial Considerations

Peer-to-peer connections, such as fiber optics, provide acceptable throughput, but at a prohibitively high cost per station. You can only determine what's best suited for your environment after analyzing the movement of work through your shop. If you listen to rumor, you might feel that networks are still far too slow to accommodate the needs of a service provider. We find that this isn't as true as it was, say, three years ago. Software such as Runshare (which dramatically improves the speed at which packets of data are moved through Ethernet cabling), as well as 100baseT, fiber optics (such as the QuickRing technology being espoused by Apple), and other advances are rapidly improving network throughput. They might never be "fast enough" in some people's eyes, but they are already fast enough to accomodate most work—if integrated and managed properly.

The capacity of your network is a critical factor here, and one that you must track carefully. Ask a network manager what she needs and almost invariably she'll respond by telling you about the dire need for more servers, bigger servers, more RAM, and accelerator boards. You don't know if you need a bigger server until you've determined how much more the one you already have can handle. We find that many times, bad server performance is a routing and workflow issue—not a lack of horsepower.

Ultimately, it comes down to cost-per-megabyte of storage. If you work on a lot of pickups—or want to expand your marketplace into segments that commonly rely on pickups, such as catalog publishers—then the off-line state becomes increasingly less effective. If a significant percentage of your work is based on pickups, then moving away from off-line for anything but long-term storage (for legal and not workflow-related reasons) is to your advantage.

Resource Considerations

There are many options available for near-line storage, with most centering around some sort of removable media. As you know, such media comes in a wide variety of sizes, but the only thing important to you is that you consider your clients' systems before deciding which media is appropriate for you. Flopticals (3½-inch opticals) are fast and sexy, but if all your clients use full-size opticals to send in their work, then it makes sense for you to consider the same devices for your own internal use.

This also reduces the types of media on which a project might be located and reduces your internal hardware support burden. Naturally you must consider speed of access and reliability, but opticals and CD-ROMs are—like everything else—only going to get faster and more reliable as time goes on.

Observations

Simplify, simplify, simplify. Don't run out and spend a jillion dollars on a new Novell system unless careful analysis has indicated the need for its powerful and extensive functions. There is always a price for features, and it can usually be measured in the cost of support and maintenance. If you're doing a lot of pickups, it might make more sense to simply buy every client an optical (or 20 opticals, if that's what it takes) than to impose a complex data management burden on everyone in the workflow. That's not to say that a comprehensive server strategy isn't important—you'll find several separate discussions in this chapter dedicated to this very issue.

What we are saying is that the processing of incoming client files, the temporary movement of work in progress to accommodate for emergencies, archiving for pickups on repetitive projects, and legal requirements you face concerning long-term storage of client work are all closely related issues that should be addressed consistently.

This isn't about technology—it's about business. Space costs you money but also presents opportunities to offer specialized services to your clients. You don't charge them for storage—but you can certainly charge them for retrieval.

PIP

25

A Formal Approach to Information Services

Abstract

How formally do you approach systems management? Accounting systems aside, do you maintain a structured approach to information management—as it relates to production workflows?

Strategic Approach

Workflow isn't a database program and it isn't a data collection effort. It's what you do with the data that's important, and any database or data management strategy is only as good as the reports you're able to generate. It's important to recognize that how you move data is a critical function, and that this can be the foundation for automating individual processes or subprocesses. However, Mac or PC gurus often stand in as data management personnel: configuring systems, rebuilding hard drives, and generally troubleshooting everyone's systems. They don't often have time left to analyze and anticipate server needs, traffic loads, acceptance ratios, and a host of other data that might prove helpful to an automation effort.

Financial Considerations

The last thing you probably want to hear is that you need to go out and add yet another computer-head to your staff. Fortunately, you probably don't need to do that.

Think about who you call upon to fix a corrupt hard drive. It's probably someone in the output department, where most of the skilled and experienced system people tend to congregate. Are there any changes in the way you process work there that might free up one of those people for full-time data work? If your preflighting structure is weak, we would bet anything that jobs spend lots of time being looked at and fixed by your most costly employees. Structuring upstream preflighting and file analysis provides output experts with very tight instructions on how to prepare for output. Such structure alone frees cumulative hours, often enough to represent an output employee's full-time position.

Resource Considerations

Designing a fireproof structure clearly costs less time and money than putting out fires and repairing the smoke damage—yet that's exactly how many sites approach data management. "Should we buy a new server? The one we have right now is awfully slow." How do you know it's slow? Has anyone presented you with actual throughput analysis? Probably not. Many purchasing decisions are based on a vague feeling that the equipment salesperson is right—we need more equipment because data isn't moving fast enough. If you don't know how fast it's moving, or what needs to be tracked, how do you know it can't move faster simply by reorganizing the way the drives are partitioned, or the way someone saves a file?

Observations

You need to determine what you want to track. Do you want to track components on a job-by-job basis? Do you want to track images separately from their jobs, relating them back to a specific invoice or job number only when the file is being output? Do you have a numbering and tracking system that fits the way your site works?

Having someone pay attention to the network only when it's broken isn't nearly the same as having someone think about data management all the time. Consider reorganizing your staff so that someone is scheduled to pay attention to data management regularly. For instance, you could schedule her time so that the first four hours of every day are dedicated to information management. That way she'll have the time and resources available to her to develop strategies specific to your workflow. She might even find the time to program your databases or workflow scripts once you've determined which processes are candidates for an automation effort.

There are many powerful tools available to you that can aid in traffic and flow analysis. The purchase and use of such tools can be a tremendous boost to your strategic efforts.

PIP

26

Commonality of Processes

ABSTRACT

When you sit down to plan a workflow strategy, make sure that you recognize the difference between "typical" and "atypical" work. You can't make valid capital equipment plans or devise workflow strategies to support a small percentage of your work. First, find profitable workflows based on experience and goals. Then, find clients with work that match the workflows. And finally, make sure your equipment list supports the workflows and the clients. To do this you'll need to identify unprofitable workflows and discard them—along with the clients that require them. This approach sounds hard-core, but it is the ultimate reality of workflow planning. Some work will never be profitable unless you drastically change the profile of your business. Other work can be remarkably profitable—even if it's not right now.

STRATEGIC APPROACH

We've all heard the system salesperson say "our workstation can process a 40MB image in real time. That other 'toy' you're considering can't do that in its dreams." That's a very powerful argument—if you routinely process 40MB images.

If, however, 90% of your work load is made up of 4MB images, configuring your system to handle 10% of your work load (the 40MB jobs) will undoubtedly cost you more than 10% of the cost to process the 4MB jobs, every single time. This is so important when considering the implementation of new technologies (direct-to-plate or digital printing, for instance) that we cannot stress it enough.

Categorize your work load. Make sure that any strategies you develop address the largest percentage of work you perform. If a specific type of job represents more than 60% of your work, then design your workflows around those jobs—even if those workflows clearly don't fit the other 40%. Substantive productivity improvement in six out of ten projects will pay for attempts to improve the remainder. On the other hand, you may have to fire some clients that now seem terribly important to your work load.

Financial Considerations

Many people buy more equipment than they really need to do the work they have—or will ever grow into. It's comforting to think that your new imagesetter has enough throughput capacity to handle three shops the size of yours. But, how smart is it? Not very.

The business world is a competitive jungle where only the quick and lean live to prosper. The reason someone buys more equipment than they need is usually because the marketing person from the manufacturer cast aspersions on equipment more suited to the actual work flow. Everyone has a primal fear of being underequipped. Well, almost everybody. Owners running the most profitable shops in the country purposely attempt to remain underequipped—but just barely.

You shouldn't have to buy new equipment—except to meet new needs. There are plenty of reasons to spend your money looming on the horizon—digital printing, proofing, direct-to-plate, and even content with cross-purposes. Wait around a little while and that little devil on your shoulder that's saying, "Buy! Spend! Buy new toys!" will have plenty of fodder.

The bottom line is that (aside from meeting minimum quality requirements) there are many ways to upgrade your performance without buying even more iron.

Resource Considerations

Generally, you should buy less equipment than you think you need. Universally, you should buy less equipment than imagesetter salespeople think you need.

Observations

Before the days of desktop imaging systems, there were really only three or four major players providing high-end scanners and workstations to the producer. Each had developed robust upgrade paths for their equipment, in keeping with its costly nature. You could count on the fact that your equipment vendor would continue to develop upgrades, since upgrade income provided a measurable percentage of the vendor's gross. Today, the world is a different place. The percentage of dollars spent on research to upgrade existing equipment, compared to money spent developing entirely new equipment, is much lower than it once was. There's an "arms race" of sorts going on between the major equipment manufacturers. It costs them, and you, tons of money to stay on the battlefield under these conditions.

Seeking commonality of processes is the key to standardization and process control—and that's where you should look first.

PIP 27

Print Servers

Abstract

Once a file has been preflighted and repaired, what happens next? Does the person doing the repair send the file to the output device and pick it up when it's done? Do people sit around waiting for things to print before they begin the next project? If so, the use of a print server has the potential to dramatically reduce "dead time."

Strategic Approach

A print server is a system that acts as a traffic cop standing between individual workstations and all the output devices that you own. It is a dedicated device—meaning that it doesn't provide storage for projects, and doesn't double as an image retouching station when it's not doing its thing as traffic controller. It serves the printing function—and nothing else.

Print servers act as the foundation for OPI strategies. OPI (also called APR by Scitex and other things by other developers) stands for Open Prepress Interface. It is a strategy whereby images are scanned and saved in a multipart format—usually four parts of a file represent the CMYK plates, and the fifth, a low-resolution version of the file, acts as a proxy used during the page makeup process. This allows for the storage of the high-resolution portion (the large-sized file) in one place, and reduces the burden of moving so much data around while the job is being designed and moved back and forth between the creator and producer sites and, no less important, within the producer site itself. When the file is RIPped, the high-res components are automatically sought and linked to the mechanical, discarding the proxy FPO (for position only) version used for the design.

Another way that a print server functions is to simply route projects to the printers to which it has access. Software such as Adobe Color Central or Archetype InterSep provide both OPI capabilities and simple routing functions in the same package. It only makes sense for software applications to approach these two functions together, since OPI and routing files are so closely related. If you don't need OPI, Color Central has a companion program called Print Central that contains only the routing applications.

If you aren't currently making use of OPI referencing, we strongly suggest that you explore its possibilities. OPI has dramatic positive effects on network traffic and allows a level of reporting and error management that's difficult to achieve using more manually assembled methods.

Routing (and spooling) work to print servers can be a very effective way to manage workflow. Let's say you have three roughly equal imagesetters, four large-format laser printers, one letter-size laser printer, two Fiery-based Canon CLCs, and a dye-sublimation device thrown in for good measure. Routing software can reduce this list of eleven devices to five categories.

An operator, when selecting an output device would only see Imagesetter, Large-Format Laser, Letter-Laser, EFI Color, and Dye Sub. Which of the three imagesetters or four lasers would actually be doing the output would be determined by the routing software, based on accessibility, which media is loaded, orientation of the page, and similar attributes. Additional routing controls allow for changes in the order in which jobs print, when they print, and other specifications such as orientation and paper source.

Financial Considerations

A print server can be built from a relatively simple workstation, and doesn't require nearly the power that a design station or color retouching station might need. Since this is really a simple routing process, it can be done on remarkably slow machines and still provide a lot of the benefits. But if you're doing both OPI and routing on the same system, the OPI processes add considerably to the system overhead. Therefore, you'll need a bigger system to support both functions. In the case of a less complex print server, you'll have considerable room for upgrades, since adding RAM and hard drives is simply a matter of plugging them in. A modest Macintosh, 486 PC (or, for the more chip-minded among us, a UNIX) system often meet print server requirements quite well.

Resource Considerations

A print server is a separate device from an image or file server and requires its own special brand of maintenance. As your organization grows and you continue to seek improvements in processing time, you'll need to dedicate a person to information management improvement (see Chapter 6, "The Database," for more information). Dedicated print servers are at the output end of everyone's workstations, and as such present the potential for too many people being involved in workstation configuration and maintenance.

Even if you can't afford to have a full-time (or part-time) person dedicated to data management, you'll need to assign the task of managing the print server to as few people as possible. Whether your structure is team- or department-based, the fewer people with access to timing, queuing, and prioritization of jobs, the better your print server will run.

OBSERVATIONS

When OPI strategies were new, they were difficult to implement—largely because the RIPping process itself was so strewn with bottlenecks and errors that routing projects often hid problems until it was too late to fix them. In today's environment, many jobs print as expected—and assuming that they were preflighted properly, routing to an active print server not only frees the operator from waiting for output, but eliminates the need for a person to guess which printer will come back on line the quickest. Automating this process alone is well worth the cost and labor involved in purchasing, setting up, and maintaining a dedicated print server.

We're somewhat surprised that more people aren't using print servers (and, to a lesser degree, OPI) to facilitate the management of network traffic and to better organize the output processes.

PIP 28

File Servers

ABSTRACT

File servers fall into two categories: multipurpose and dedicated. Most client/server network strategies are based on a multipurpose approach, where all components are stored on a central file server. Many manufacturing sites—particularly those where OPI workflows are in use—benefit from a data management strategy that puts images on one server and the remaining components that make up work in progress on another.

Strategic Approach

There are several issues to contend with when designing or implementing network strategies. First and foremost is the question of whether or not the network needs to actually facilitate file transfer or will simply be a storage device. When an individual needs to work on a project, are the requisite files moved to that person's local drive? Workflow and throughput often dictate that they are. Disk-intensive work in particular is far better executed on the device connected to the SCSI port than one connected via Ethernet. Retouching and color correction are clearly in this category, and often require that images be moved from the server to the local drive for the duration of the intimate color work before being routed back to the server. In this case, larger local drives with shuttles to facilitate movement might prove faster, cheaper, and easier to manage. If scanning systems are producing high-res OPI images, then a dedicated image server makes a lot of sense.

Financial Considerations

From the standpoint of data collection as it relates to workflow events, job tracking, time and throughput analysis, and similar functions, having a centralized file server is invaluable. Sitting in the middle of the factory, it serves to gather everyone's input and organizes such data in a manner that eases the production of reports.

As a centralized storage strategy, a file server also is invaluable for staging and preflighting incoming jobs. Since staging and preflighting are more a matter of looking at (as opposed to working with) files, it isn't a memory-intensive operation. A file server purchased for the customer service department and dedicated to the staging and receipt of incoming work can prove a very sound investment.

When the time comes for a more skilled operator to actually open and repair the file, the choice exists to do the work on the file as it sits on the server, or to move the file to a local drive. This decision really needs to be based on your particular situation. There's no question that working with a local file is far faster than working on a file that's living on a server.

You are challenged to determine which method of access is more effective—working from the server or moving the file to a local drive. Both have advantages and disadvantages that can only be weighed on site relative to a specific job. We will say that many highly successful shops use file servers and removable media (also called sneakernet) to solve specific transfer problems.

Resource Considerations

Every time you add a server, it increases the complexity of your network. Servers must be backed up continually; this should be based on inexpensive and stable media—such as DAT or similar tape devices. Servers must be maintained as well—moving lots of files on and off the drive fragments the disk after a relatively short time and requires that the disks be formatted occasionally to maximize their speed and throughput.

Again, we come back to the importance of an at least partially dedicated staff member whose responsibility it is to pay attention to these details. If your server needs are complex, someone must be responsible for maintaining close and intimate communication with the vendor that sold you the servers.

Observations

Are applications being run from a central server? Except for database applications, few of the tools used in a typical production workflow will operate in the conventional client/server mode. In the past, many computers worked this way—you bought one copy of an application and a license for each user. The same application ran on everyone's workstation. Today's more modular, object-oriented software is clearly designed to run in a one-person, one-application environment. Therefore, any server strategy has more to do with storage and centralization of data than it does with spreading resources over more than one workstation.

RAID drive technologies are increasingly popular components in server designs. They ensure a level of data integrity that's tough to argue against. They cost (considerably) more than regular drives, but the difference will be made up the first time your system doesn't crash—or crashes and comes back up with a simple restart.

PIP

29

The RIP

ABSTRACT

The RIP in your system is the heart of your manufacturing workflows. The choices you make regarding your RIP can have dramatic effects—both good and bad—on your shop's throughput. In our opinion, the most important aspect of the RIP lies in your ability to upgrade its software.

STRATEGIC APPROACH

When choosing a RIP, you face two basic considerations: whether to rely on software- or hardware-based machines. Hardware RIPs are currently more common than software devices. Software-based RIPs normally run on a dedicated workstation—either Mac-, Windows-, or UNIX-based. Clearly it's easier to upgrade a software RIP than it is to upgrade a hardware RIP, since the latter upgrade often requires a trade-in (more accurately, a trade-up). Hardware RIPs must be upgraded by the developer; software RIPs, on the other hand, can be improved simply by adding RAM to the RIP server, more hard drive space, or even by swapping its workstation for a more powerful version. And with a software RIP, you can buy the RIP from one source and the marking engines from another. (Imagesetter manufacturers, by the way, currently make considerable money selling new hardware RIPs and might not like the sound of this, but we stand by it.)

A software RIP—actually a collection of software applications—is easily upgraded by purchasing the upgrade (usually on a CD-ROM) and running a simple installation program.

Generally speaking, if you're in the market for a new RIP, consider moving to a software-based RIP.

Financial Considerations

Dollar for dollar, there isn't much cost difference between the two types of RIPs initially. A powerful workstation designed to run a software RIP can easily reach $10,000 or more and, when added to the cost of the software, will probably equal or even exceed the cost of a hardware RIP. In the long run, though, the software RIP will be easier and cheaper to upgrade. And since you can improve the speed and performance of a software RIP by purchasing widely available RAM, storage, or accelerator boards, your ability to comparison shop is greatly enhanced.

Resource Considerations

Hardware devices are basically "plug-and-play" in nature. Software RIPs do, however, require a fair amount of maintenance. A software RIP has to be installed on a workstation where the data management overhead is considerable. And if you buy a RIP that runs on a UNIX or even a Windows machine, and your staff's expertise is limited to the Mac platform, the human-resources issue must be resolved. UNIX workstations are very "keyboard intensive"—to run a UNIX RIP is considerably more difficult than running a RIP in a Windows or Mac environment. Despite this, UNIX provides true multitasking functionality and more easily accommodates the move in our industry toward using multiple RIPs to break complex work into manageable chunks.

Observations

Since RIPping is so important, it requires constant attention. Assume you will need to upgrade your RIP even before the upgrades are announced.

As workstations gain in horsepower, we will see more and more software-based RIPs. RISC technology, once the sole domain of high-end workstation manufacturers, is rapidly becoming a standard processor model for computers such as the Mac. It won't be long before all popular computers—Intel-based as well as Motorola designs—will be RISC-based. We anticipate that RISC-aware software RIPs will eventually dominate the field.

Also consider that software RIPs are more suitable for customization. Increasingly, software RIPs will be customized through extension technology for specific industrial needs such as packaging, textiles, and even multimedia applications.

One last consideration that affects the viability of software RIPs is the proofing cycle. As digital presses and direct-to-plate systems become prevalent, the concept of outputting pages to several devices—from the exact same RIP—will become very important. Since different RIPs produce different results, converting to an entirely digital output environment necessitates using the same RIP to proof that you use to output.

PIP 30

Imposition

ABSTRACT

Imposition is a highly variable, device-specific event. In most cases, imposition decisions should remain flexible until the last possible moment. Digital imposition is a stripper's worst nightmare. While perfect for some environments, it is equally inappropriate for others. Make sure you know which category your shop falls into before you spend too much time or money on digital imposition applications.

STRATEGIC APPROACH

The first thing you have to ask yourself is if the type of work you routinely process is suitable for digital imposition—not all work is. Books are best suited to digital imposition. After that, any multiple-page signature is a candidate, but the complexity of the files and your success rate at getting those files from preflighting to perfect film should be primary factors in determining if imposing film digitally is viable. Assume that you can develop hybrid workflows to auto-impose some flats, while leaving others for manual processes. Try and get your strippers to buy into the digital process—their input is invaluable and (please note) they will not necessarily be replaced by automation. You will always need a stripper's skills to design and maintain the imposition process. Imposition software isn't smart—at least not yet. It still needs an operator to tell it what to do.

Digitally stripping flats has a clear economic advantage over doing it manually—again, assuming you know the job will output properly. If you're having problems outputting single forms, then chances are you're going to have eight times the headache in a digital imposition workflow. Intimately related to the direct-to-plate process, digital imposition will eventually emerge as the standard output method in commercial printing environments. It would be wise to buy into the learning curve—if not the equipment itself.

Financial Considerations

Ask yourself if buying a (very) large-format imagesetter is more cost-effective than the purchase of a step-and-repeat camera. There are many on the market, and years of shop experience indicates they work perfectly well. And there are many high-volume environments using large-format cameras that are quite productive. But, while very popular, they do require a lot of human intervention and depend largely on a manual assembly process—at least for base images.

Make no mistake about it, a large-format output device for outputting large imposed forms is a costly proposition. You're looking at well over $200,000 to take the plunge (the software isn't cheap either—starting around $4,000).

Resource Considerations

If your shop runs smoothly and has controlled workflows, introducing digital imposition will not be a resource drain on your staff and equipment. You should, though, consider the time it takes to RIP an eight-up signature. Realistically, the digital process—without any manual stripping—is considerably faster than manual or hybrid workflows that require skilled laborers working with razor blades and glue.

Running imposition software requires a solid knowledge of specific press conditions within your factory. Beware: your techies have a natural desire to control all things technical—especially if it concerns software that might run on their pet systems. Does it make sense to turn over the intricacies of digital imposition to your file-repair experts? Instead, have your strippers and press-savvy production pros learn the imposition software. If they've never used a Mac or PC, it isn't too difficult to teach them how. The Mac guru, on the other hand, probably doesn't know the first thing about work-in-turn. All the computer knowledge in the world isn't going to teach them. It's far more productive to teach a stripper how to use a computer than it is to teach a chip-head about imposition. Don't let anyone tell you otherwise.

The best people to train for digital imposition understand conventional imposition. If they can design conventional forms, they should be able to design digital signatures that meet your requirements. Digital imposition is device-specific, but you might need the flexibility to change presses just before a job runs—this demands agility in the process of signature design. Software packages that provide imposition functionality also allow you to design and store custom signatures that could prove to be beneficial for your regular clients.

Observations

An ideal workflow in a shop where digital imposition is being used would look like this: individual forms are received from the client and preflighted; after any necessary repairs or collections, the file is imposed; OPI referencing is applied to the entire flat; traps are made, now or right after the file is converted to CMYK; after conversion, the file is output to an "impo" setter (an eight- or four-page imposition imagesetter); and finally, the form is proofed, either digitally or with laminate media.

PIP

31

Telecommunications

ABSTRACT

Modems have a reputation for being too slow to accommodate high-resolution files. But there are many options available to the manufacturer or creator who wants to take advantage of electronic file transfer. Digital, high-speed telecommunications is now becoming affordable and easier to use, making the passing of files from creator to producer, and back again, increasingly easy.

STRATEGIC APPROACH

There are three electronic transfer methods to consider (well, actually two, and we'll tell you why). The first is a standard modem. Today's modems are relatively fast, with common transfer rates running from 9,600 baud to 28,800 baud. Without getting too technical, let's just say that 9,600 is fine if you want to send E-mail to someone, but fails miserably beyond that. At 28,800 baud, you can transfer a 1MB file in about six minutes—that's still far too slow for a 40MB page.

Conventional modems convert digital signals to analog noise and pipe it across conventional phone lines (called POTS, for Plain Old Telephone Service). In large part, that's why they're so slow.

The other two transfer options use digital lines; in these instances, data is sent from one location to another without getting converted to analog—it stays in a digital format. There are two primary options for digital lines.

The first is called ISDN and is available just about anywhere there is a phone company. ISDN lines are two specially-engineered POTS lines that are "bonded" together. They're designed to transfer digital data. (Note: they can double as a voice line with an optional converter. Another option, called Switched 56, is actually half of a full-width ISDN.) Using just one of the two lines for transferring digital information (and the other for a dedicated fax, let's say) provides speeds up to approximately

64,000 baud. If you use them solely for file transfer (as opposed to also accessing the Internet or commercial on-line services), you can use both lines simultaneously and achieve throughput of around 40MB per hour—not too shabby.

The second and fastest option is known as TX lines (where X stands for 1, 3, or even 12). A standard T1 line uses 24 individual analog (POTS) lines wired together in much the same way as the ISDN option. You can send over 200MB an hour, maybe more, using this technology. Newer iterations (T3 and higher) can double, and even triple, that speed. And you can also configure a TX line so that some of the lines are used for your regular voice phone system and some are dedicated to data transfer.

Financial Considerations

ISDN lines are fairly common now and many companies already have them. It costs about $100 to have the phone company hook up an ISDN line, about $500 for a special digital modem, and another $60 per month to maintain the line. If you consider the difference in speed over conventional modems, ISDNs are pretty cheap. You might consider installing them for your clients in lieu of a discount. Since you can also use an ISDN line to access the World Wide Web, the value to your client is considerable. (Add about $30 a month to the costs outlined above for 30 hours of Internet access, and you'll really have your clients dancing on their desks).

Tx lines, on the other hand, aren't as cheap or as flexible. There's a price to pay for speed—about $5,000 to set up (per site). And a TX line only connects with other TX lines, so both sides have to take on the initial financial burden (not so with ISDN, although the transfer speed is limited to the slowest modem in the line of transfer). So you can see, even though we said there were three options, TX lines might be too expensive for most small and mid-size shops to consider.

Resource Considerations

Once installed, high-speed data lines become essentially a "drag-and-drop" application, and really don't require much to maintain. Setting up the lines at first is another story altogether—unfortunately, your phone company won't do much more than drop the line somewhere on your property. In most cases, too, the phone company does not provide technical support. You'd better count on hiring outside help to get you started if you don't have an expert on staff who's done it before. Setting up a TX system takes a lot of trial and error to get all the "switches" right (both software and hardware). Rather than let your own staff learn as they go, it's a lot more cost-effective to bring in third-party assistance. Simply communicating with the phone company about telecommunications is an art form in itself. Hire a specialist if you have any doubts.

OBSERVATIONS

Having high-speed lines in your shop not only shows your clients that you consider technology important, but they can save time by transferring files back and forth between your shop and their offices. Could they be faster? Sure, but even at current speeds, the added value goes far beyond convenience.

Having high-speed lines gives you and your staff workable access to the World Wide Web. Considering how many companies are already using the Web as an advertising and promotional venue, it might benefit your company to do the same. But be careful! Don't be seduced by the Web! A recent Seybold conference focused almost entirely on the Internet, to the exclusion of other issues that might be making us money today. The Internet is not about to replace prepress and printing, especially as a revenue source.

PIP

32

Calibration

Abstract

In today's industry, there are two important aspects concerning calibration. The first concerns standardization of input and output between you and your clients' sites; the second has to do with linearization of the equipment in your shop.

Strategic Approach

The "portability" of color definitions remains one of the most difficult issues in the graphics arts. Hope is on the horizon, however, in the form of the International Color Consortium (ICC). The intent of the ICC is that everyone who builds or designs scanners, monitors, output devices, and proofing systems will develop a "profile" of their equipment that will allow color to be moved from one piece of equipment to another without any appreciable degradation or distortion.

How often do you calibrate the devices in your shop? In the past, scanner operators routinely pulled random proofs of known tonal values in order to determine how far from "factory defaults" a specific device was at any given time. This ensured that the device remained "linearized" or adjusted to meet factory defaults. Does linearization occur at your shop at regular intervals? If not, consider starting right away. It's likely that many of your devices are out of line.

We know there are a lot of calibration devices on the market, but we do wonder about claims that suggest that a suction cup hanging from your monitor, or expensive boards that reside in just your workstation, or any of the other dozens of products that claim to provide the perfect calibration solution will in fact do what they claim.

Financial Considerations

Both Apple and Microsoft have accepted ICC color standards and have allowed space within their operating systems—known as independent color space—for color transformations. The software on the Mac side is called ColorSync™. ColorSync isn't calibration software per se—it's merely the location in the operating system where comparisons between source and output devices are made. It doesn't really cost any money to institute calibration within a trade environment; it simply requires that your core software library is savvy to the technology. In other words, sit around for a year or so and you'll see color become quite predictable in the digital workflow.

Resource Considerations

Some software lets you calibrate input and output color models. Photoshop, for instance, provides several methods of linearizing its internal data structures with external devices, such as monitors and output devices. For example, the EFI software for Photoshop does a great job displaying on-screen what will be output from a Canon color copier. As color profiles continue to be developed by software and hardware manufacturers, such cross-application calibration tools will become more functional, and therefore more popular.

Keep in mind that any software or hardware that attempts to calibrate (or more accurately, match) profiles between disparate devices must depend on one of two technologies. The first is algorithmic. It calculates on the fly while data is moving from one place to another (from scanner to monitor, for example). Since the computer (or "firmware") literally makes up the figures on which to base its color transformation from one device to another, algorithmic calibration is not as accurate as it might be.

The second approach uses color lookup tables and is far more accurate, but a bit cumbersome. A CLUT (as they're called) is literally a library of the entire color reproduction gamut available from each specific device. The libraries are quite large and take a lot of time to create. A 24-bit color lookup table is a whopping 64MB and can take about 45 minutes to create for a scanner, monitor, or output device.

OBSERVATIONS

When considering calibration, there are a number of issues that should concern you. First, each device in your workflow must be linearized and its actual output or input adjusted, or calibrated, to match factory specifications. This ensures balanced neutrals and noncasted tones. In a large-scale environment, equipment should be linearized and calibrated at least twice a day.

The other issue to contend with is the conversion of RGB data to CMYK data. Despite its critical role in transformations from one color model to another, this conversion is increasingly falling on the shoulders of mere humans running highly variable software packages. To consistently achieve high-quality image reproduction, it still takes a great deal of skill and experience. No impending technological development is likely to change this. Traditionally, CMYK conversion occurred in hardware concurrent with the input/output task. Now this conversion must be executed, and indeed calibrated, to your specific output conditions.

In the immediate future, we're likely to see an increasing number of clients bringing in RGB data files that appeared to be OK on their monitors but lack essential tonal information—such as the location of specular and diffuse highlights and identified shadow values. The reason this happens is that performing a "mode change" within Photoshop—from RGB to CMYK, for example—is not as accurate, or repeatable, for image reproduction as more sophisticated software available on high-end systems is. If you are not very experienced in reproducing color, then consider developing a workflow using specialized software that isolates and controls when and how data is converted.

We should note that several sites we know are using Photoshop for very high-quality color work. It's not impossible to do—just difficult to master if you are not experienced in color reproduction already.

PIP

33

Job Bags

ABSTRACT

What's a job bag today, anyway? The traditional definition describes a physical container that holds the components of a project as it moves through a shop. It often has forms printed on it to ease the movement and gathering of data relating to the project. An electronic job bag has far more potential for tracking a project than the physical job bag did.

STRATEGIC APPROACH

How you define "job bag" and the way you approach data management are closely related. An electronic job bag can be a database that tracks project-specific information: job number; descriptive information such as number of pages, trim sizes, special instructions; and a host of data that used to be written on the outside of the physical bag you used before. Additionally, an electronic job bag can provide calculations based on time and dates to facilitate the measurement of a job's flow and latency. (See Chapter 6, "The Database," for more information.)

In a digital environment the location and movement of items in the job bag is critical to a job's progress. Copying files is one example. When you work with a specific client logo, for instance, do you make copies of it every time it's needed, and put each copy into a folder containing all the other project components? While this creates multiple copies of the original (and opens the door to revision problems), it also eliminates any possibility of linking errors that may have been introduced, and dramatically reduces network traffic. Rather than "hit" the server every time the logo is required, the server is accessed only when the job starts and provides the necessary piece of artwork. This can all be automated and verified in-line to a great degree, through a proper image-tracking database design.

Financial Considerations

In a discussion about electronic job bags, there are two cost-related factors at play. The first is the cost of servers—if you pursue any form of digital job bag, it will function best if run from a centralized server, where data entry functions are structured to update a single database.

The second expense—and it can quickly grow out of proportion to the expected results—is programming. Custom programming is a viper's nest of potential disagreement, missed communication, incorrect perceptions and expectations, and lots of other bad things. You have to be very careful in setting down exactly what it is you want and what you expect to get in return for your money and time.

This suggests that you should stick to simple, "flat" databases that require little, if any, custom programming. The more complex the database, the more likely it is that programming that database will prove problematic.

Resource Considerations

You'll need a programmer, or at least someone in the organization who knows how to use a program like Claris FileMaker Pro. Since it operates on both Macintosh and Windows platforms, FileMaker Pro provides a relatively rich construction environment, with every type of field you'll ever need.

Additionally, FileMaker Pro is "script savvy," meaning that it can run and/or communicate with external applications. Using a workflow controller such as OPEN, or Mainstream from Agfa, a good programmer could conceivably develop an entirely electronic job bag.

Observations

Do not attempt to create electronic job bags until you have done an exhaustive analysis of your process map and have determined exactly what types of data belong in the bag. Additionally, you will have to match your job-bag strategy to the way your network is configured. Always think in terms of reducing network traffic—moving large data files around your shop often creates a jam.

PIP

34

Architectural and Environmental Considerations

ABSTRACT

Is your plant designed to facilitate the movement of work, or is it designed to compartmentalize functions? How does your plant look when clients are invited in for a tour? Is it a jewel of efficiency that reflects your commitment to excellence—or is it more like Times Square on New Year's Eve?

STRATEGIC APPROACH

Workflow, while digital in many respects, still requires the physical movement of elements. Your success is also largely dependent on verbal and visual communication between the individuals working for you. Walls do not encourage communication, nor does the strict isolation of one department from another. In many ways, using production teams helps alleviate this problem, but it isn't a substitute for a well-thought-out floor plan.

The customer service department, for example, is often placed away from the factory and near administrative functions. While this model may work for some, it makes sense to explore alternatives. Putting customer service closer to where the jobs are actually done puts the CSR much closer to the work for which he or she is responsible, thus encouraging communication with the folks on the shop floor. It is easier to ask someone a question when that person is 10 feet away than it is to fill out a report that might get delivered at the end of the month. Other benefits from proximity include "creative overhearing," as mentioned previously. And part of cross-training is seeing what other job categories do all day, especially how they interact with your function. How you use the space you have can impact how business is done.

Financial Considerations

Changing the physical arrangement of your business can be approached in many different ways. We know plants that were actually built from the ground up to facilitate communication among employees. A case in point is Central Florida Press (a member of the World Color group) located in Orlando. President Bob Brach often boasts of having "the longest hall in Florida." He's probably right. Reaching hundreds of feet, the hall is a great example of architectural planning.

All of the offices and rooms on the east side of the hall are devoted to administration, conference space, and employee break rooms. The west side of the hall is the factory. It's designed so that paper and ink arrive and are stored on the south side, and the Post Office and shipping facilities are to the north. In between are the presses. Ink is piped throughout the shop in color-coordinated tubing. The shop is spotlessly clean and, since they don't use alcohol (even though Heidelberg originally insisted that they couldn't get by without it), amazingly odor-free.

Obviously, not everyone is going to build their own factory from the ground up—but it does show the lengths to which some managers and owners are willing to go to match physical space to profitable workflows. On a more realistic level, start by simply discarding partitions and moving a few desks around—this won't cost you a dime. At Baum Printing in Philadelphia, Pennsylvania, they literally broke down walls in an attempt to "metaphorically" break down walls between departments.

Resource Considerations

How would you describe your shop?

Clean and bright? "Tours move through the shop with ease and everyone pays attention to their work space. Each department or team room is well-organized and open. There aren't a lot of walls, but we do pay attention to privacy and security. People looking at us feel confident that we take pride in our work, and that we'll treat them like professionals."

Dark and dingy, with old cheap paneling, grimy from years

of vaporized ink and chemistry? "It takes us days to get ready for a tour, and invariably there's an animal carcass or a stack of grand jury indictments laying in plain sight. It's impossible to get people to clean up after themselves—they're a bunch of animals. The refrigerator in the break room has been declared a biohazard by the EPA. We're thinking of applying for status as a Superfund site."

Need we say more? How your shop looks is indicative of more than having an efficient cleaning crew. It's a barometer against which you can measure how you and your staff think of yourselves—and even more, what visitors and other potential income sources think of you.

Observations

Very rarely do we meet successful owners and managers who don't constantly consider how they look and how well their environment reflects, and facilitates, the type of work they do.

Study the movement of work through your shop. Take a floor plan of your building and draw a line that shows where a project goes from the time it comes in the front door until it is shipped out the back. We've done this for clients, and very often the resultant map looks like a psychotic's directions to the River Styx. Carefully analyze what needs to be done to the project, where the various processes take place, who's involved in those processes, where their personal space is located, and how you might move things around to make jobs flow better and life a little easier.

Making life easier means reducing the complexity of movement. Try to avoid moving components in opposite directions; otherwise, someone or something will need to retrace steps.

From a workflow standpoint, your goal should be to simplify movement. From a business standpoint, try to maintain the most professional atmosphere possible. This doesn't mean glitz—it means you should look like you're ready to get down to business. Encourage and motivate your staff to take pride in the surroundings—and be sure to practice what you preach.

PIP

35

Facilities Management (Outsourcing)

Abstract

As new markets and opportunities present themselves, consider initially outsourcing new kinds of work, such as digital proofing or stochastic screening, to determine its viability as a service you want to provide.

Strategic Approach

If you study the early days of desktop, you'll find that many large trade shops, when faced with the need to output PostScript mechanicals, initially bought the services from service bureaus. This approach allowed the trade shops to test the waters, so to speak, before plunging in. When it proved necessary to make PostScript page processing part of their workflow, some simply bought the service bureaus they used. We don't suggest that anyone can do this, but trying out new technologies with someone else's name on the bottom line is an acceptable way to research new services.

Financial Considerations

Buying a service from someone else is often more expensive than it would be to produce the service yourself. (There are some things, such as bindery, foil stamping, aqueous coatings, and other functions that many commercial printers continue to, and probably should, purchase "outside.") Whether or not to buy others' services to try them out, or because you don't want to do that kind of work yourself, is a decision you'll have to make relative to your own business.

Resource Considerations

If you choose to provide new services, you'll need two things: first, a client base that requires the type of product you're considering; and second, a desire to provide the product or service. This isn't quite as simple as it might sound. An excellent example of what we're talking about is high-speed, on-demand print services, perhaps now at copier, rather than ink-on-paper, quality. If you're a high-end color trade shop or commercial printer, your clients are spending at least as much on this lower-cost work as they are on the high-end output you print or produce for them. You might say, "So what? I don't want to be a copy shop." This is all well and good if we weren't facing the day when high-speed copying and quality sheetfed start to look the same. By then it might be too late to run out and change your business plan. On the other hand, if you were to solicit this type of business now, and make arrangements with a high-quality copy business to handle the load, then your potential vulnerability will be considerably lower in the future.

OBSERVATIONS

A modern-day proverb says, "Why buy it if you can rent it?" High-speed duplication or multiple-color copies might not be in your future, but they're certainly in your clients' future. Rather than making so dramatic an addition to your product list using your own (perhaps limited) resources, why not take advantage of the fact that there are already people out there with the equipment required? This is an increasingly common tactic with color digital printing-press work, where investors in the technology are eager to have others resell their services.

An offer of expanded services, when made to a receptive audience, can prove to be an indicator of market potential. If you make a somewhat formal introduction (seminar, brochure, direct contact) of a new product or service, you may very well be overwhelmed (or, alternatively, underwhelmed) by the response. Regardless, whether it's bindery, high-speed copying, postal services, multimedia design and production, or myriad other new product lines, you can use your clients and someone else's workflow to test your idea. While brokering other firms' work may seem like heresy, it has a long history in the graphic arts market.

PIP

36

A Plan for Continual Improvement

ABSTRACT

If you want your business to grow, you need to know how big you want it to get. Continuous improvement depends on knowing where you are today and where you want to be tomorrow. While increasing profits overall is a long-term goal, there should also be measurable short-term goals to track incremental progress, particularly in areas where production workers have some control. Short-term goals, when strongly rooted in long-term ones, are strong motivators for your employees.

STRATEGIC APPROACH

You don't have to be a rabid TQM disciple to understand that developing a plan for continuous improvement will make it easier to improve workflows and heighten profitability. The biggest problem is where to get started. Try first to identify areas that provide accurate numerical feedback that will help focus your improvement efforts. These numbers ought to give you a clear picture of how well current methods are working. This is far better than having some vague, idealistic plan for overall improvement.

On the other hand, beware of collecting useless numbers. For instance, of what benefit is knowing the number of key clicks in the word processing pool, or tracking time spent in the rest room? This isn't a classic time-and-motion study. You aren't trying to reduce your employees to automatons. Rather, you want to help them improve procedures and efficiency overall. And remember, faster isn't always better. If an area of business is functioning well, adjusting how the work is done, rather than doing it faster, may produce better results in the long run. After taking a close look at how you do your work, you will probably find that most of the problems are not due to the failings of individuals, but to the inadequacy of procedures generally.

Among the key areas that measure productivity are: the ratio between hours worked and hours billed; the quantity of rework; on-time delivery; and wastage and spoilage of film.

Smaller landmarks can also be tracked. One shop keeps track of the number of times it sends jobs using special overnight air couriers (sometimes even counter-to-counter) rather than normal air couriers. The extra charges for super air service add up very quickly and are tied directly to an inability to meet deadlines efficiently. If these kinds of charges plague your business, meeting schedules needs to be improved by altering the workflow. Shipping during normal business hours will save you a bundle.

It is critical that all your number trackings be made public, at least within the plant. Public posting of the latest numbers should be a matter of pride for the group and the plant as a whole. Once you have reached maximum improvement in one area, start looking for others to track.

Financial Considerations

The cost of tracking continuous improvement, once sound data-tracking systems and regular employee meetings are in place, is fairly minimal. It does, though, require the timely assembly of numbers into a report. Meetings cost something in terms of time, but needn't be an overwhelming burden if you have an agenda and stick to it. And getting good suggestions from motivated employees may save you the cost of a consultant.

Resource Considerations

You need to assign someone the duty of generating the reports or charts and posting the new numbers. You also need bulletin boards in areas easily accessible to employees. And remember, celebrate heartily when improvement milestones are reached.

Observations

Think of continuous improvement in terms of primary and secondary goals. The first goal of every employee is to deliver assigned work on time, at appropriate quality levels or better.

The second goal—one that goes beyond the day-to-day—is to find ways to improve or streamline the first task. Again, the improvement methods will best come from the people who do the work.

Getting employees involved in an improvement plan creates a strong sense of teamwork that they can each be proud of. Tracking performance numbers is a way of keeping score. Most employees want to contribute to the greater good of the organization and do more than collect their paychecks, especially if they feel that their input is valued. This means not just doing the work, but figuring out how to do the work better, faster, or less expensively.

While you may get dramatic improvements at first, the later improvements will come more slowly and perhaps be accompanied by occasional steps backward. Don't try to fix everything at once.

CHAPTER 6

THE DATABASE

Every time an event occurs in your shop, data is generated. The question you really must ask yourself is, "Which data is important to track, and which information is not?" We're dismayed by the time and effort expended to enter, track, and manage information that has nothing to do with making money—or even counting money, for that matter.

So let's take a look at data collection and management from a somewhat different perspective. We'll also depart from the commonly-held belief that database programs somehow form the basis of a workflow strategy. In our opinion, this is way off the mark. Software isn't workflow—and workflow isn't a software application.

Workflow strategies are just that—strategies. They are built on management-related, process-related, or people-related decisions that are not produced by a specific hardware device or software package. While strategic decisions generate a need to track specific information in a defined manner, they aren't based on software applications, data-entry screens, or centralized file servers. Workflow improvement efforts are based on your ability to lead and motivate, as well as your skills at communicating the needs of the organization to clients, staff, and peers. While it may not be the essence of workflow, accurate, sensible data collection and analysis are tactically important to any workflow reengineering plans you devise. That said, let's explore the issue of database management.

Collecting and managing data takes time. Since you're a service provider, not a programming company, data management can consume considerable amounts of time and resources you probably don't have. This isn't to say that such efforts aren't important—just that you should always weigh the time you spend against the potential return in terms of eventual productivity gains. You can sink many hours and dollars tinkering with database programs—proceed with caution and an eye toward simplification.

There are two aspects of data to consider: the first is the collection and organization of the data; the second, storage of the data. It shouldn't surprise you that you're already dealing with data management on both levels. In your accounting department, collecting information forms the basis for the endless reports accounting personnel seem to generate. The second category, storage, refers to the management and movement of project components, such as mechanicals, high-resolution scans, line art, imposition information, color lookup tables, and the like. The latter data represents the files that you're processing.

The former represents data you're collecting about the state and attributes of components being effected by those processes.

Why Collect Information in the First Place?

While there are several perfectly good reasons to track certain data, you have to ask yourself which information is important to your workflow. If you expect to operate at a 90% or better productivity level, you really don't have time to count or track information that you aren't going to use later. That's our first recommendation: don't track something for its own sake.

The trick is to pass pertinent information—along with pages and page components—to team members further down the workflow. When Person A finishes her part of a workflow, she passes it to Person B. Person B needs certain data from A in order to move forward with his specific task. This type of internal data tracking and reporting facilitates progress.

Imagine getting a file with no documented information. "What is this I'm looking at? Is it a TIFF or an EPS file? Has it had unsharp masking applied? Was it corrected or just scanned and stored?" Without pertinent data, the person asking these questions must spend time poking around on the disk drive to see what file types are in the folder. Determining if a file has been corrected or sharpened would be a little more difficult and would require that they go talk to someone. They might have to visit another part of the plant, or send a query over the e-mail system, but they can find out—assuming Person A hasn't gone out to lunch with Person C or gotten the image confused with another one. You get the point.

This type of data "sneakernet" seems to occupy half the industry—especially in an unstructured workflow. A database—assuming it is capable of reporting exactly the right amount of information (and no more), and is also capable of delivering it to whoever needs it at the exact moment it's required—has the potential to dramatically improve efficiency and therefore speed up the workflow, while cutting down on misassumptions and misinformation.

Moving component-specific information down the pipeline along with the elements in question is the primary reason to consider database technology as an aid to developing profitable workflows.

Keep in mind that such internal or component-specific information must be presented to the reader clearly and concisely. Supply them with too much information, or make it difficult to grasp quickly, and you've defeated the purpose of tracking the data.

A second argument for using database technology is to provide yourself and your co-workers with valid statistical, profit-related information. While component-specific data is tracked to facilitate the workflow itself, profit-related data tells you whether or not your workflow strategies are producing the desired results. You can set up the most sophisticated data-tracking system on the planet, but if you have no baseline from which to measure its impact, you might as well have spent the time playing CD games. Statistics, while potentially dull and dry, remain the only way to measure success. And by the way—if statistics are showing improved profits, they don't seem nearly as boring.

Many owners and managers—perhaps yourself—run their business partially by the numbers, but more often by the seat of their pants. They rely on guesswork and intuition. Before we get in too deep, let us say that this approach has worked—and worked brilliantly—for many sites. The downside is that it only works long-term if you are very lucky or a prescient genius. Since we consider ourselves neither very lucky nor abnormally intelligent, we feel that most people can gain control over their profits only by knowing where they are right now and where they want to be in the future. Numbers work. They give you a baseline and keep the target in front of you (and your staff) at all times.

Where Do I Start?

A prime reason to track data is to help you identify areas for improvement. Clearly, there's far too much information moving around to view at one time. Once you've determined what kind of data you want to see, you have a starting point. From a business standpoint, the only reason to spend a dime on developing a database is to generate solid, effective, and—most important—readable reports. It's the report that you should be focusing on, not the data-entry screens, not the horsepower of the software, not its ability to easily format dates, and not the cost or popularity of the application. Just as software isn't workflow, workflow isn't a database. Clear, concise reports should be your goal.

Reports are a significant, rhetorical arrangement of data. (By rhetorical, we mean they tell a clear and persuasive story.) Facts themselves really don't mean anything unless they provide a framework in which to analyze a problem or make a decision. Reports should help you think. If you were to go to your system right now and secure a "dump" of invoices, amounts, dates, client information, line items, and delivery instructions, it would be of little help. The only thing such an exercise would accomplish would be to encourage you to further sort and gather even more data—"When were these generated? What was the gross profit in each? What does all this stuff mean?"

The efficacy of any report is highly dependent on how the information is organized. A perfect example would be a report that showed profits on a job-by-job basis (see next page).

August Invoice Report

Client	Invoice	Amount	Profit	As %
Johnson Co.	19034	$12,104	$809	6.6%
Harrison Lumber	20332	8,322	2,200	26.4
Fractal Motors	18332	34,522	14,437	41.8
Sophie's Wigs	20354	18,982	-1807	-0.09

As a productivity tool, this is clearly better than an invoice listing that failed to reflect the profit on each invoice. It could, however, be a lot *more* useful if you were to consider the arrangement of the information. For example, there's data you could compare from month to month. How much more valid would August's information be if you also had July's data in the same report (see below)?

Profit Tracking Report

Client	July	Profit	August	Profit
Johnson Co.	$14,932	22.9%	$12,104	6.6%
Harrison Lumber	8,460	27.0	8,322	26.4
Fractal Motors	5,233	21.1	34,522	41.8
Sophie's Wigs	21,933	-1.3	18,982	-0.09

Same data, different result—especially as it relates to your ability to identify workflow problems and improve your profits. While the first report showed that Fractal Motors represented your most profitable client and Sophie's Wigs was a real bomb, it didn't help you to identify trends. And there are some interesting trends happening here. This is valid data—not different, mind you, just *shown* differently by making use of proximity to improve the usefulness of the report. The first report surely provides you with data—but the second provides you with data you can *act* upon.

Let's start with the Johnson Co. account. Here you see that

between July and August, the same volume of work resulted in profits that fell from almost 23% to less than 7%. You can bet the problem happened on the shop floor somewhere. Either the job profile forced it to move through an unprofitable workflow (or, worse, there aren't any workflows defined and every month brings with it some new experiment), or something caused major rework, or someone dropped the ball.

The problem might, of course, be upstream. Maybe the client hired a new production manager, for example, or hired a temp to lay out the magazine. This report lets you know you have a problem—right now. If you had a structured workflow plan in place, you would even know where to start looking—and what to look for.

On the other hand, the Fractal Motors account somehow went from good to wonderful in the same time frame. Why?

Did someone think of a way to more effectively process the job? Was a different workflow tried for the first time? Did someone work extra hard that month? Or extra smart? Whatever happened, you should find out and try doing the same thing with the Sophie's Wig account—maybe you can turn it into an acceptable project instead of doing what seems to be the best course of action: firing the client, the salesperson, or both.

If you can hit 30% profit on one client, there's no reason you can't hit 20% on all clients—that is, if you continually seek to match the ideal client profile to functional and profitable workflows. It isn't magic, it's management—and your ability to communicate to your salespeople the importance of finding profitable clients. If you use databases intelligently, they can provide you with the proof you need to convince your salespeople that some clients are just not profitable.

Design your reports first and "reverse-engineer" your prospective database to meet your needs. *Then* get someone to fiddle around with a data management or spreadsheet program to collect the data you need to make the reports, so that you end up analyzing data that's worth analyzing—data that helps you to think and is presented in a rational and simple-to-understand manner. Designing and (quietly) delivering these kinds of reports are the reasons you're paying those bean counters, isn't it? (Don't tell them we called them that—it really annoys them.)

Defining a Database

What is a database? It's two things.

First, it refers to the actual software program or language that's used to create fields, records, indices, data-entry screens, and reports. There are many different products from which to choose, and they range from simple and inexpensive, to incredibly complex and costly.

Second, a database refers to how the software manages and collates the data it contains. In theory, a "flat" database is one that can manage individual collections of information—a personnel file, a recipe book, an image database, an invoice listing, or a project database. A "relational" database allows you to collect information from one record (let's say a master client file) and use it to automatically enter information in fields in another database—such as an invoice database. While entering information on an invoice screen, you could type in a client code, and the program would go out to the master client file and pull in ("relate") the client's address and shipping information. Pointers—small programs that use access codes to point or locate—look at key fields in one file (the client code, in this case) and relate them to information found in another. Another way of looking at relational databases is to imagine a report whose every column (or row) might display information from a different file.

In reality, there's little difference between the two, and don't let the comparison take too much of your time. Claris FileMaker, considered a flat database, offers extensive opportunity to combine the contents of disparate databases into a single report or screen. FoxPro and Oracle, touted as relational databases, do essentially the same thing, but provide a more extensive "language" with which a programmer might establish those relations.

It's important for you to realize that the program your staff uses to create the database isn't the database itself. By our definition, the database is a collection of records specific to your business and workflows. The program used to define the fields, records, indices, entry screens, or reports doesn't know—or particularly care—what information it's tracking.

Data collection

The next issue is how to collect information and gather it into a useful database. There are a few methods for doing so.

The first is to automate the gathering of data. The most popular approach in the Macintosh environment is based on AppleScript, an object-oriented "scripting" language that is quite simple to master. It can actually "talk" to and "hear" various applications, including the Finder or core operating system. It can send information to an application, or receive feedback concerning the results of functions that the software applied to a job component. For example, AppleScript can have a FileMaker database find out from Photoshop the color model of a particular file. Subsequently, it could pass that information down to another application—such as Color Central. Color Central would then know the color model, and with that information it could act accordingly—such as raising a red flag that says, "This image has to be converted to CMYK and sent back to me so I can do the OPI referencing."

Scripting relies on system-level technology known as Apple Events. Developers are increasingly designing their software (and "firmware") to be Event-savvy—but not all software fits this model. AppleScript is very limited in what it can do if it doesn't have access to the core functionality of a specific program. When

Apple Events
Apple Events is a messaging language used by some applications to communicate with other applications. It allows programs to share data and commands. An Event, for instance, may wake an idle application, causing it to perform some task or process some data.

it does, however, it has tremendous power in automating data entry. This makes AppleScript particularly well-suited for gathering and using the type of data that moves alongside digital pages (and the elements that make them up).

The second method of data entry can be thought of as being driven by external events. This includes bar-code readers, keypads, and information that's written on a form or entered into a data screen. This type of data entry often happens at the moment an event occurs: "I've stripped the job. It's ready to go to Cromalin proofs." The person responsible clicked a button or passed a bar-code reader over the job ticket when it arrived at his station, and clicked, stroked, or pressed the appropriate button when the task was completed. The data is fed back to a central database.

Many products purporting to be the end-all of workflow strategies are actually this type of data-collection tool. Examples abound in our industry, and we certainly don't mean to cast aspersions on the considerable development efforts that go into designing and creating them. But—we repeat—they aren't workflow. If you consider the collection (or entry) of data as a subprocess in the development of an overall workflow strategy, you will see these programs for what they really are—highly effective gatherers.

The next method involves batch-entry of information. An example of this can be found by looking at your accounting system. Payables packages usually batch all the checks you've written into a single posting, and pass that collective information to the general-ledger module. There it's used to establish specific line items or is entered in the appropriate account categories. When all is said and done, you (hopefully) have a useful set of financial data. Batch-entry is highly suitable for collecting statistical, management-focused data. Again, concentrate on collecting data that will improve profits, not spend time. Don't collect or massage data for its own sake.

For example, in the scenario we just discussed, when the

stripper got the job and when he or she clicked the button to indicate that it was finished might not be very important if your shop is running smoothly and routinely garnering 30% profit from anything it touches. If, however, jobs always spend too much time in stripping, it will become evident quickly—and you probably won't need a detailed report to let you know. Some things—such as a poorly thought-out procedure that is a sink of time and effort—can be identified without a major, costly programming effort.

Always look at data collection as something that doesn't really make you money. You need data, granted, but new requirements (passing the wand, pressing the key, calling up a data-entry screen, etc.) take time to do. Make sure the return justifies the cost—and don't assume the cost can always be measured in dollars. Always attempt to get the most bang out of the least effort. Designing the report first (or defining the information required to automate a process) and then determining how it should best be gathered is usually an effective approach.

Reports Come in Many Shapes and Sizes

There are many different forms that reports might take. If you follow our advice and begin any database strategy by first designing a report, it helps to identify how you might format the ones you need.

The first kind of report is typified by a job listing. It might be sorted by invoice number, data, alphabetically by client, or even by volume, but most are made up of a set of events or components. They're usually listed horizontally across the page.

The second type of report is more sort-dependent, and is columnar in nature. A good example of this type of report would be the profitability model we showed you earlier in this chapter. Other examples would be reports that showed how many jobs from Fractal Motors are in the shop at this point, how many of them require rework, how many require 200-line screens, how many fall below an acceptable profit margin, how many proofs we need to do today—the list of what might be important to you is probably quite lengthy.

Next come analytical reports, within which rudimentary calculations are performed on the numbers found in the various columns. These calculations might rely on specific, predetermined figures (such as what you consider to be acceptable margins) or on variables found in other columns—or even in other databases. Again referring to the reports we discussed earlier in this chapter (see page 192), the "As %" column represents this sort of data and how it might be used to build a useful report. A logical extension of this would be another column that indicated (perhaps with an asterisk) whenever a margin fell below, let's say, 25%. (By the way, if you think the types of margins we're tossing about here are totally unrealistic, it's a good thing you're reading this book.) Either way, raw dumps of numeric information are rarely useful without such on-the-fly calculations.

Don't hesitate to keep asking your accounting people to give you something new, or something formatted differently, or to design and produce composite reports like the ones in our example. They're not too busy to do that—no matter what they say. The more profit you can find and retain, the more likely they are to retain their positions. If you don't have money to count, you don't need anyone to count it. Getting to the point where your reports are flawless, on time, and 100% useful takes time and effort. Don't count on getting it right the first time.

Once you've determined how you want your reports to look, you can begin to name them. Naming them is important. The name must reflect the purpose—much the same as a job description should match the requirements you expect a person to meet.

Reports don't only have to present data for your analysis—they can be used to raise flags when something isn't quite right. An example would be an exception report. Such a listing would show events, projects, or processes that—for some reason—failed to meet your requirements, or required special attention. Examples of the types of information you might want in such a report would be: how many jobs took longer than two days? What jobs required rework in the past month? What jobs fell below an acceptable profit margin? How many jobs were refused on first showing? You get the idea.

Finally, consider the effectiveness of presenting statistical data in a graphical manner. Charts are wonderful things—they can summarize a million numbers and turn them into a line. Depending on what you're tracking, the direction of the line indicates the direction things are heading in your factory. At Lanman Lithotech, there were profitability and rework charts at every single station in the building. At a glance, everyone in the plant could see how changes in process controls and workflow strategies continued to improve profits—and tighten workflow. One line (profits) went up (a good thing), while another (rework) went down (also good). Other lines indicated productivity gains (and losses) on team and company levels.

Make sure that your charts are as useful as your reports—beautiful pie charts, for example, fail to provide you with very useful comparison data—and comparisons between similar events that happen at different times form the basis of good charts.

Creating the Database

Perhaps you want to design your own proprietary database engine that gathers data like a demon and spits out the greatest reports in the world. You probably should start with a very complex database program that requires an experienced, highly-paid consultant to design, create, debug, and maintain. Sound like fun?

If, on the other hand, you want to build your database strategy on a solid report generator, consider Claris FileMaker Pro.

It runs on both the Macintosh and Windows platforms. It's simple to learn, powerful, and completely AppleScript savvy. We think it's a safe and solid choice. While there are many programs to choose from, FileMaker is the most popular data management application in our business.

Before purchasing a database be sure it is useable by normal human beings. Even the most intuitive database languages require a good deal of programming expertise to handle the IF/THENs, GOTOs, ELSEs, and other programese. If you keep it simple, it's conceivable that you could simply pick up a program and design your own reports.

Another means of access for normal mortals is a query language. Queries can be as simple as custom-designed screens that pass search criteria on to the main data file. Examples include a screen that allows a CSR to access progress information about the scans in one of his jobs. Even clients could use such a tool to track their projects.

Whichever way you go—complex or simple—designing a database isn't the domain of programmers. Why? Because the database must—we repeat, must—be designed by the people who need the reports and who must act upon the data being tracked. If we had to identify one problem that always happens when a producer hires a programmer, it's that he fails to properly convey exactly his expectations to the programmer. Instead, the programmer is supplied with vague specifications, and the work begins. The programmer can't define your needs on the fly while also writing code for buttons, screens, and reports. Set the pace and the specifications—then find someone to execute your desires. And make sure you dry-run any data-collection strategy before committing it to disk.

Finally, design reports and entry screens that are visually well-organized and attractive. You would be surprised at how much this affects people's willingness and ability to enter accurate information.

The Image Database

Managing and monitoring the images that circulate through your factory is one of the most critical functions of a database system. Here are some important data categories.

Image Database

Field	Data Type/Description	Optimal Collection Opportunity	Subsequent Events Requiring Collected Data	Producer or Creator Specific
File Name	A system-level name assigned at the time a file is first created. It can be assigned by an operator, or created by the addition of a suffix affixed to the existing name following a specific event. Streamline, for example, adds the suffix .ART; Acrobat the suffix .PDF.	The moment of file creation	Any event requiring the use of, or reference to, the file in question	Creator and Producer
Location	Where the file resides on the server	When first created; whenever file is moved; during any Save or Save As event	Any event that needs to incorporate or directly affect the image	Creator and Producer
Size	Physical disk requirement for storing the file	When first created; whenever file is moved; during any Save or Save As event	Events requiring assignment of disk space; transfer of file from one station to another	Creator and Producer
Resolution	The width and height of a file, expressed in terms of pixels.	When first created; from any resize or resampling event	Trapping events, OPI referencing, imposition; rendering to marking engine; re-purpose to alternative delivery vehicle	Creator
Source	Where the image came from. This will eventually be very important when all devices are "profiled."	At the input event	Any event requiring transformation from one color space to another (i.e., RGB to CMYK; YCC to RGB; etc.)	Producer
OPI Referencing Data (could be more than one field)	1. Data providing pointers from FPO/samples to high-resolution originals. 2. Data connecting attributes such as rotation, sizing, location on the page, etc., that might have been applied to sample during creative event(s).	At input; following input during separate sampling/distribution event;during specific design-specific event (i.e, when an image is moved, rotated, or sized within the context of the layout application)	OPI referencing, which in an ideal workflow would occur sometime immediately following the imposition event	Producer
Audit Trails of Specific Functions	Information that could recreate previous ontological or environmental conditions.	Native preference conditions; responses to dialog-box inquiries; file headers; separate indices; reference libraries such as are used in the Photo CD inter-polation strategies	Events relating to post-process quality control or fine-tuning.	Producer
Time and Date	The time the file was created, and the time the file was last modified. These are extremely important in determining completion times, which file represents the latest changes, etc.	Operating-system-level function, read by or entered into field when document is created or any time it's resaved	Project planning	Creator and Producer

Let's analyze some of the information in the table on page 201.

FILE NAME. The name of the file is important for obvious reasons: without a name, you can't find the file. This isn't to say that there aren't a lot of files being created invisibly by native applications that the operator never sees or knows about. There are times when these "invisible" files might carry important data, and might therefore need to be accommodated in any attempt to automate the process or entry. You might also consider an internal and external name: one that the designer assigned such as "Saturday's Child," and one that matches your internal requirements, which might include the job code, client, date, and processes completed.

LOCATION. You always need to know where the files are. Information concerning the location of a given image is important for OPI referencing.

SIZE. Clearly, you can't process a 3GB project if you have only 30MB available on your server. When we discuss the project database, we'll see how the total cumulative size of imported files will be derived from the contents of this field.

RESOLUTION. This is the optimal amount of data "in the hole," so to speak. Effective workflows rely on the relationship between the resolution of a given image and the output requirements (line screen). Too much data slows things down, and too little data results in inferior reproduction.

SOURCE. Since many color management strategies rely on the reproduction gamut of specific input and output devices, the source of the image will become increasingly important. Whether the color management system relies on device profiles or color lookup tables, it needs to know which one to use.

OPI referencing data. In certain cases OPI workflows provide the absolute best of all worlds to the creator/producer environment. Catalogs, for example, benefit dramatically from the designer's use of low-resolution placement proxies linked to high-resolution originals that are stored on your servers. More important, though, from the standpoint of future workflows, OPI referencing will serve a much larger role: facilitating the automation of page assembly within the context of the producer site.

Audit trail of specific functions. Since there are so many things an operator (or process) might do to affect the data in an image, it's important to track what's been done already. For example, you never want to resize an illustration after you trap it, or retouch an image before you correct it (doing it backwards often causes artifacts to appear).

Time and date. A time and date field is derived directly from the operating system. Since the OS routinely tracks creation and modification dates and times, this information is available in real time. It's vital to the workflow process, especially when considering automation (for more information, see Chapter 7, "Automating Workflows"). Reports based on time sensitivity rely totally on accurate time keeping. Ensuring that a file is the latest revision is also possible through use of these system-based fields. Apple Event-savvy programs access OS-level information, including creation and modification dates.

There are a few other important items to consider in the data collection process as well.

Where a file is being used. This information would prove highly useful within both the creator and producer environments. It could be used, for example, in any archiving event that sought to determine if a file should be kept on-line. In the creative environment, knowing where a file is being used could help identify which file(s) were used on a specific campaign, or for a certain client. This information is stored within the context of a page layout file; for instance, you can see which images have been placed in a Quark XPress document by checking under the Utilities menu. You can't, however, look at the original files and tell whether they've been placed in currently active documents—or which documents those might be. Related information might include how many times an image was used, and if it exists in other forms (say, an RGB version, a YCC version, and a CMYK version).

WHAT EVENT IS LIKELY TO COME NEXT. For example, if an image is placed inside a page layout or illustration program, and that file is destined for distribution as a sheetfed page, then clearly a transform into CMYK color space is in its future. An event might be triggered to move the file into a space reserved for this to happen.

Any image database should also be designed to track information specific to vector art such as Adobe Illustrator drawings.

FLATNESS. A PostScript operator that controls the number of data points utilized to describe a given curve. The smaller the number, the fewer the data points. In the case of very small numbers, a curve could be reduced to a series of straight lines, resulting in a very jagged curve. Higher smoothness factors result in much smoother curves, but require proportionately greater memory and processing horsepower. Complex images take more time to RIP (and cost more money) than less complex drawings—even though they might look exactly the same when output.

CUSTOM COLORS. Most drawing programs allow for the creation of custom colors, which might be composed of process builds, or might provide ancillary data about custom colorants, such as spot colors; varnish specifications; metallics, flourescents, or other custom colorants; die-cut shapes for embossing, foil stamping, or cutting; etc.

TRAPPING INFORMATION. This is a tough issue to resolve. If a designer is building traps into the actual creative process, there's a great chance that they'll miss something, or that the techniques aren't readily available to meet specific trapping requirements (ignoring white, trapping two blends, or choking the three primaries when white type is reversed out of a rich black, for example). When this happens, it's very difficult to design a process whereby built-in traps can somehow be stripped from the drawing. When is a specific choice for a rule color or object definition part of the cognitive design process, and when is it just an attempt at trapping the element in question?

THE NUMBER OF OBJECTS IN THE DRAWING. The more objects a drawing contains, the more pressure it places on the RIP. Before a specific RIP is selected, attention to this factor might influence the decision about which device to use or when to schedule it.

MASKS. The attributes associated with masks created within a drawing (or image editing programs, for that matter) can prove vitally important in environments where files might be moving back and forth between proprietary and open-architecture systems, such as between a Scitex and a Macintosh.

NESTED EPS. Information relative to the use or existence of embedded EPS files. Another major workflow blockage is the use of multilayered EPS files, which, by the way, can also present a major challenge to the design of an automated production process. As software becomes more sophisticated, components from different source applications can be mixed together; for instance, designers are using more and more photographic elements within the context of their illustrations. Elements created in Photoshop, for example, are often imported into masks and other elements within an Illustrator drawing. This creates a nest of EPS files, each calling data from the other. When the resultant illustration is output independently, the problem isn't so apparent. But if you place that illustration into a page layout program, then the fun really begins. Heaven forbid that the drawing also contains fonts, or that the drawing contains yet another placed drawing that contains fonts! Fonts within embedded EPS files present one of the most common causes of rework in the production environment.

FONTS. Drawings often contain typefaces, whether independent or within the context of embedded EPS files. Which fonts have been used in a document, and whether or not those fonts are available on the system doing the output, stops or slows down more jobs than any single event. Someone, or something, had better be tracking fonts—those used in a specific drawing (or document) and those available on a system at a specific moment in time. (This is another area that benefits from anticipation of subsequent requirements; you know darn well that when that file moves to another location, the fonts had better be with the file.) Although technically illegal under terms of most font licenses, clients make copies of the fonts and send them along to the service provider—this is the currently accepted way of dealing with this data requirement. An automated workflow environment would deal with this in some acceptable (to the font vendor as well as within the context of cost-effective page processing) manner.

The Project Database

The function of the project database is to manage all of the disparate components that make up a given job. The project database tells you what jobs are in progress, what their requirements and specifications are, and which files are necessary to achieve the desired results.

Database fields might include:

PROJECT NUMBER OR NAME. This identifier relates all objects in the content list. In an open, unstructured environment (one where no database is actively or passively collecting key information), when placing an object on a page, or importing an element into a drawing, or creating nested EPS files, only the highest object on the "stack" knows the identity of elements contained within its structure. As database technology evolves, it's apparent that some accommodation should be made to tell the imported objects that they're being used—even if the file containing them isn't open at the time.

A CONTENT LISTING. This field could size itself dynamically to accommodate an inventory of all the elements of the project in question. This might include: a simple list of the elements, and sizes, locations, resolutions, curves, etc.; and any information deemed to be important to subsequent events, contained within the context of individual records, as defined above. The content listing would essentially gather, combine, calculate, sort, and then report—either on paper or digitally to another application—on specific fields contained in individual database records.

SIZE. The cumulative size (in megabytes) of the project, collected from the content list.

CURRENT STATUS. Given that a predefined process would provide a list of events that would have to be completed before the next event could occur, a field in this database could conceivably tell you how many of those events are done at the moment of the inquiry; that is, a percentage of completion.

PROBLEMS. In an automated-preflighting environment, the database might respond with a list of what errors or problems were encountered.

AUDIT OF EVENTS. This field accommodates compression events, transforms, curve adjustments, key-plate (black) calculations, and other information that needs to follow the project along the workflow pipeline.

Naturally, you'll have to design your own databases based on your specific needs. We've provided a starting point for your own efforts. It's likely you've been wrestling with data management for years—most businesspeople have. It isn't necessary to trash everything you have and start all over. You can, however, take a fresh look at how you use the data you already have.

Relationships between Different Types of Data

The place your data management is most developed is within your accounting department. Since accountants and financial officers already work with data-capture systems—based on generally accepted accounting principles—they don't have to be quite as imaginative as, let's say, your graphics system manager.

Do you really need two or more people managing your data? Perhaps you don't, but in most cases, conventional information management people approach data management from exactly the position that we warned you to avoid. They love tracking data for its own sake. To a certain extent, so does your accounting department. In a perfect world, you could assign the creation and management of image and project databases to the same person who manages your accounting databases, but in practice you're probably better off with someone who understands both the manufacturing processes and how to use FileMaker or similar applications.

The argument that conventional database skills don't necessarily migrate effectively to the shop floor underscores the need to choose a package that's friendly enough for almost anyone to use. This argument also relates to your ability to effectively link data between different processes within manufacturing, as well as between production and accounting.

Fortunately, most popular accounting packages accept standard ASCII tab- or comma-delimited records—and any solid database engine like FileMaker or FoxPro can generate both types of data.

If your accounting system is already generating reports that are at least partially useable at this point, it's best to focus your initial database efforts at getting the production side of your data management up to speed. There are two issues to deal with: component data and statistical data. Production-specific statistical data needs to be drawn from, and integrated with, other accounting information. Figures derived from invoicing, estimating, cost accounting, sales and marketing expenses, general overhead, return-on-investment calculations, and a host of other financial data—when placed in proximity to production-specific data—take on a whole new meaning.

Make sure you keep the lines of communication between your accounting people and your production people (especially the ones working with the database design) wide open. Don't settle for reports that provide you with less information than you need. It's not so hard to move columns around, find averages, or generate trend lines. Motivate your technical personnel as you would motivate production personnel—use positive reinforcement and instill a shared sense of importance for everyone's role in the big picture.

Another issue to contend with is how and why you might let people outside the plant gain access to various parts of your system. There are two groups to consider: your clients and your sales staff.

Client Access

Letting clients into your system offers many advantages and only a few pitfalls. The use of high-speed telecommunications for exchanging information and files is rapidly becoming not only a workflow issue but a competitive one as well. Many service providers are packaging and selling, if you will, the ability to transmit projects with free on-line services, access to bulletin boards, and other related services. It's sexy for the client, can actually work if it's done right, and might be a deciding factor when a publisher or agency is considering a new service provider. If two vendors offer similar quality and service, and one comes with an ISDN line and access to their servers, it will win every time.

A database system capable of providing progress information could also provide that information to a client, if the database was designed to allow external access. This is the application we described earlier, when we talked about designing query screens—these are small programs or screen designs that act as a window into certain data on your system. A client could conceivably double-click on an icon on their desktop and be able to see where their work is, and how it's progressing. As long as you have your act together (that is, they don't find out the job is going to be late), this accessibility can be a very strong motivation for using you instead of the guy down the street. It is a value-added service in its truest sense—you design the system to give yourself the tools to improve your workflows, and the same data is also valuable to your clients. Kill two birds with one hard drive, so to speak.

Another type of database that you could share with your clients is the support database. Documents concerning service, training issues, technical "how-tos," tips and tricks, and techniques specific to your manufacturing methods can be designed and put on your server. A small front-end shell (friendly entry screens) could be designed in a program such as FileMaker, as can query screens for distribution to your clients (or prospects). With a few clicks, they could find out how to handle traps, set resolutions, determine trim and bleed characteristics, and a host of similar information that is of interest to the client and simultaneously supports your profitable workflows. Clearly, this isn't as high a priority as production and statistical databases are, but a support database has a lot of appeal for the creative community.

The creation of this type of service may be purely nonproductive time, though. Keep that in mind whenever you are considering fun projects like this one. Fun projects can be costly projects. That's not to say there's not marketing value here, but remember to set your priorities, with productivity and profits as the barometers of success.

Sales Access

Salespeople need to access the database system when they're on the road, as does any employee on the road whose assignment requires information residing at the plant. Both Windows and the Mac OS offer several options for remote access. The concept is fairly simple and well within the scope of your database design. And it doesn't cost a lot to add this capability—except, of course, for the portable computer your salesperson would need. Portables and laptops have declined in price quite a bit—now, a useful machine costs well under $3,000. If your sales force is on the road a lot, or if you have remote locations, consider the addition of access software and perhaps even a partial subsidy for their personal machines.

Security

Once you begin to think about letting people have access to your system, the issue of security raises its ugly head. In reality, security isn't limited to your clients or people surfing the Web. It also includes vulnerability to sabotage from within.

We know companies that employ full-time locksmiths—so they can change door and entry accesses in the blink of an eye. While we certainly don't suggest that you go this far, you should consider security of data an important aspect of your database management strategy. Workflows have a strange way of grinding to a halt when important files are mysteriously deleted. Make sure that whatever your security strategy is, important access codes are not in the hands of only one person. Having different shifts be responsible for their own backups, or even passing the responsibility for data integrity down to the team level, are both options you might look at.

When someone calls your system from a remote location, the concept of a firewall comes into play. A *firewall* is a set of applications that literally block access to information below a specified level of access to your system, just like a firewall in a building inhibits the spread of a blaze. This makes it harder for someone to post your financials or your customer list on an Internet newsgroup. It doesn't make it impossible, though.

When links occur between your production databases and your accounting system, both systems are accessible—however briefly. During this time, a serious hacker could conceivably download the sensitive or mission-critical data. Not to worry though, there is a workaround.

Batching data entry limits the time both databases are exposed. If your systems are linked around the clock, you give a potential hacker (or industrial spy) more time to poke around. If, on the other hand, information is passed from production to accounting in batches, the time that one system is actually connected to another is dramatically reduced. It is during these periods of connectivity that a firewall comes into play. Software exists to create such walls, but we strongly suggest if you are going to have a situation such as the one we're describing, bring in outside help—good, trusted, reliable outside help.

There are other situations less drastic than this. Not all secure information resides on the accounting server, nor does only public information reside on the production servers. The truth is there are both types of data on both servers. The trick to ensuring security is to design outside access in such a way so that the firewall stands between the browsers and anything you don't want them to see. A good consultant is really the only sure-fire (pardon the expression) way to make sure you are covered. And don't even dream about an Internet presence without first thinking about security because the Internet is where the real hackers live. In your own system, with your own clients, the security risk is greatly reduced.

A Snapshot of Current Technology: The Lanman Lithotech Database

To draw a parallel between our theoretical discussion of databases and how they are currently being used, we spoke with Bob Nuelle, director of advanced technology at Lanman Lithotech in Orlando, Florida. Featured in this book (as well as the *1995 Apple Annual Report*), the site is one of the most advanced output environments in the country. As we discussed in Chapter 3, it provides lithographic services to some of the most demanding customers in the publishing industry, and it does so in a 100% PostScript workflow. It processes thousands of pages of high-quality color on Macintosh and Windows platforms. More germane to this discussion is the fact that it has put the database concepts we've discussed into action which has resulted in measurable improvements in performance.

Nuelle has had experience with a wide range of fully relational database languages, yet has standardized on FileMaker Pro, by Claris Corporation. Although not truly relational (in a pure sense), the program does allow for extensive one-dimensional lookups (that is, you can automatically enter information found in another database by entering data common to both records). For example, a customer code links an invoice file to the customer database, eliminating the need for the operator to enter the shipping and mailing addresses (they're stored in the customer file, and brought in as needed).

Nuelle feels that FileMaker Pro provides more than sufficient horsepower and accommodates most of his needs. He runs individual copies of the software on every machine in the building, and maintains a central server within the customer service department.

Nuelle maintains two databases. The first, an image database, contains information about every image in the shop:

- Creation or modification date
- Name (as assigned by client)
- Job number (related back to the project database)
- Folder where the image is stored
- Drive where the image is stored
- File type
- Physical size (in inches)
- Virtual size (in kilobytes)

He considers these the minimum fields for image management. Individual records might also carry additional, optional fields specific to a client's particular needs:

- Keywords
- SKU
- Bar-code (an image-based field)
- Product categories (sportswear, evening wear, etc.)

- Status (relevant to: square; silo; partial or breakout silo; actual file vs. shadow—gray shadows are stored as separate images, etc.)

Nuelle also maintains a project database. This is a fully relational, free-form collection of individual files that are maintained for a variety of related functions (customer databases, invoicing and accounting, percentage of completion, customer service requests, etc.). This database contains:

- Job number (relating all elements found in the image database)
- Customer name
- Job name
- Customer's job number

It also has hundreds of job- or project-specific fields, such as:

- Number of scans
- Scan in/out dates
- Random proofs
- Proof types
- Film at each stage
- Profiles of project types (for example film, proof, assembly, scans, archives, etc.). This is essentially the workflow map for each project being tracked.

The project database becomes, then, a project planning tool. It provides information key to the decision-making process so important to profitable and effective operations in the environment. It can generate time lines, and help manage resources, both human and device-dependent.

The relational nature of databases is the key to long-term information management because individual routines, processes, and information "buckets" can be modified or introduced without having to redesign a system in its entirety. As demands change, Nuelle plans to evolve to an Apple Events-savvy model.

Currently, the system is designed to turn data streams into a tool set for internal communications. The data becomes the vehicle for the message itself, while the message itself increases in complexity. Therefore, Lanman must continually model its databases on communications at its most basic levels. Nuelle feels strongly that the firm must eventually develop profiles for everything: every person, every piece of equipment, and every event sequence. He relies on the ability to dissect files using an event-driven analysis tool—like a highly automated database—to identify and correct problems at the earliest possible point in the workflow.

Before we move on, we want to make a final remark about the concepts we've been discussing in this chapter. The comment is drawn from an excellent recent book, *Reengineering the Corporation,* by Michael Hammer and James Champey. The authors state (and we annotate):

"With an expert system [i.e., an intelligent database], a generalist can perform the work of many experts."

CHAPTER 7

AUTOMATING WORKFLOWS

As the publishing industry approaches a point where 100% of all projects are created and processed on computers, the need within the production community to automate the collection and preparation of digital pages becomes more evident. Yet the whole idea of automating our often quirky and error-prone industry causes a smile of disbelief to come to the lips of experienced managers. No one imagines that prepress in its current state could be performed by pressing a few buttons; it will remain a labor-intensive business. But the thought of tying together some steps in a workflow is starting to look more and more possible. There is no magic recipe, but the combination of a solid workflow analysis, good data collection, and a scriptable workflow automation program is likely to be the set of tools that will allow for gradual, step-by-step automation of many functions.

Indeed, the switch from analog to digital workflows has caused serious disruptions. Not very long ago, when a project was received at the typesetter, trade shop, or commercial print site, the mechanical that had been prepared by the creator contained specific instructions—and visible components from which those instructions could easily be gathered. You could measure the page with a ruler; you could use a scaling wheel to immediately determine the size settings for any scans the design required. The components of digital mechanicals, on the other hand, aren't so easily dealt with by purely mechanical means. How can these jobs be automated?

But this confusion is a transitory problem that is already starting to clear up. Although digital jobs can be more difficult to deal with, that's only because we're still trying to deal with these digital components using analog thought processes. That doesn't work. Digital files are not only more suited to automation efforts, they must be automated if the industry is to reap the advantages they offer us. Too much focus has been placed on the flexibility that PostScript design tools provide to the front end. We've only just begun to explore the real implications of digital workflows from the standpoint of their "automatability" on the manufacturing side. Since the earliest adopters were designers, their needs were addressed first. It was only natural. When the development industry turns to servicing the needs of the manufacturing community, the designer's tools will probably make another quantum leap in their overall impact on productivity. Right now, we take digital files and force-feed them through hybrid workflows.

In the previous chapter, we presented ways that a database might facilitate the automation of production workflows. The key concept we attempt to define in this chapter is: PostScript Manufacturing (a similar term, used in other industries, is computer assisted manufacturing, or CAM). Automation, while it relies heavily on the consistency of data, is fundamentally different from a database. True automation, by definition, occurs with a minimum of operator intervention.

Automation software can be thought of as process-control software. We will build our discussion on a collection of process control tools developed by Adobe Systems known as OPEN. A similar set of tools from Agfa (a division of Bayer Corporation) is called Mainstream.

Digital Workflows

The first digital tools to make their appearance at the creator site were used in the creative process. Production sites, in the meantime, had gradually made the move to digital systems about six or seven years before the earliest desktop systems. Digital production methods were already in place at those production sites who had the economic wherewithal to make the required purchases—which often costs millions of dollars. The shops that had made the move to digital systems did so in an effort, primarily, to automate their workflows and therefore reduce the staffing required to maintain them; and secondarily, to enhance their product mix with high-end digital retouching services, therefore gaining (buying) an advantage in products and services that would equate to a better competitive position.

Without a doubt, trade shops moved to digital systems—scanners, recorders, and workstations—mostly to aid the automation process. At the same time, they ran parallel manual workflows. The ability to automate the scanning process, color-correct and adjust images, and render individual elements (loose color) was the first real indication that the parts of the production process could be automated. (These proprietary and costly systems were known as CEPS: Color Electronic Production Systems.) While a digital scanner still needed a scanner operator—one that cost more than a camera operator, since they needed a lot more training and experience—productivity improved because one person could do the work of four. Nevertheless, because of its great investment, for a long time digital workflow was an alternative to proven manual methods, which were a fallback strategy.

The desktop revolution has made the manual workflow all but obsolete. When the design community embraced doing creative work on desktop computers, their mechanicals immediately evolved into a form that was fundamentally different than those they prepared before desktop. There were only a few points in the production cycle (final page assembly, trapping, and imposition) where manual methods made sense.

But this workflow caused significant problems, as we have noted above. Individual digital components, unlike conventional, "pasted-down" page components, are not as visible as they move through the required steps. They might appear on the screen, and they might even show up on laser proofs. But each component in the electronic mechanical has unique, invisible attributes specific to its portion of the overall content list. If we know the kind of press a given page will be printed on, for example, these invisible attributes—things such as resolution, screen angles, special colors, page sizes, page count, where high-resolution images are positioned, and how the job should be imposed—are buried inside each of those components. And, as interactive and other distribution modes gain in popularity, such as multimedia, network distribution, and digital printers, the files created by the designer will increasingly carry time-sensitive information as well. Furthermore, the nature of short-run and on-demand printing technologies indicate that there is less time budgeted for human interaction with the files.

With all these changes, automation is no longer just an attempt to cut expenses by eliminating staff. It is a response to increasing demands of new technology. These include faster throughput on increasingly complex and diverse jobs, processed within greater time and budget pressures, under circumstances that allow for a wider range of errors than ever before. It is only through true automation that users of desktop technology—on both sides of the producer/creator equation—will begin to achieve the results that the technology promises. In all the years we've been using desktop systems, we haven't come close to reaching the full potential that computers promise. Computers are at their best when executing unattended tasks.

What's missing from today's workflows is a tool set that would allow you to:

- Determine and define a single event or sequence of events to be executed on a digital page (or a series of digital pages) without operator intervention. Additionally, the tools should let you store these definitions in a digital library, so that they could be called upon any time a similar document needed to go through a similar series of events.
- Edit those definitions.
- Trigger a series of events that are actually performed by individual applications.
- Move information concerning individual page components along with the working file(s), so that subsequently triggered events could utilize information generated during previously applied events.
- Monitor a nd direct ongoing sequences in real time, so that the file could be paused, stopped, flushed, etc., to meet dynamic conditions on the production shop floor.

Defining a Production Event

Today's workflows are task-oriented. In the near future, workflows will be process-oriented. Page forms will continue to stabilize, and many of the variables that now force producers to "tinker" with native file formats (Photoshop files, XPress files, PageMaker files, Photo CD files) will disappear. Distilling a file prior to its being shipped off to the producer site will become the de facto standard at the creator site.

Distillation
Distillation refers to the simplification and regularization of PostScript files. Adobe Acrobat Distiller is a type of PostScript RIP that takes in PostScript code and outputs tighter, DSC-conforming PostScript code.

Distilling, using Adobe Acrobat Distiller, simplifies a file and removes many native attributes and peculiarities, leaving only that information that is important to the processes required to complete the project at the production site. Once the PDF format matures to handle a wider variety of prepress situations, all files arriving at the producer site will be distilled PDF files—not a folder full of version-specific, OS-native-specific files as they are today. PDF distillation breaks all jobs into a series of predictable components. This is an important criteria when you want to define and execute a series of events that need to happen to that file.

Once a digital document has been designed and approved within the creator environment, the electronic pages, including any page elements used during the design, are moved to the production site. This document has been distilled into a file model that is highly predictable when it is received. The source of this new capability is dependent on software developments that are near.

But even the best software isn't enough. We mustn't overlook the ability of structured training to achieve a very high level of predictability from the creator files. If good training programs exist, the variabilities likely to be found, even in native PostScript files, can be dramatically reduced and brought under control.

Unlike the creative process, where the contents of individual page forms are very fluid and changing, the document that arrives at the production site should be eminently well-suited for automation. This requirement will only increase in the future—even as documents move toward broader content in the form of sound, motion, and interactivity. As a document moves across the "bridge" between the creator and producer site, the content must solidify, so that subsequent processes are, for all intents and purposes, automatable. In the future, the producer site will spend much more time designing and refining workflow definitions (those sequences we've mentioned) than they will opening, editing, and resaving native-application files—because automation will take care of the hands-on work.

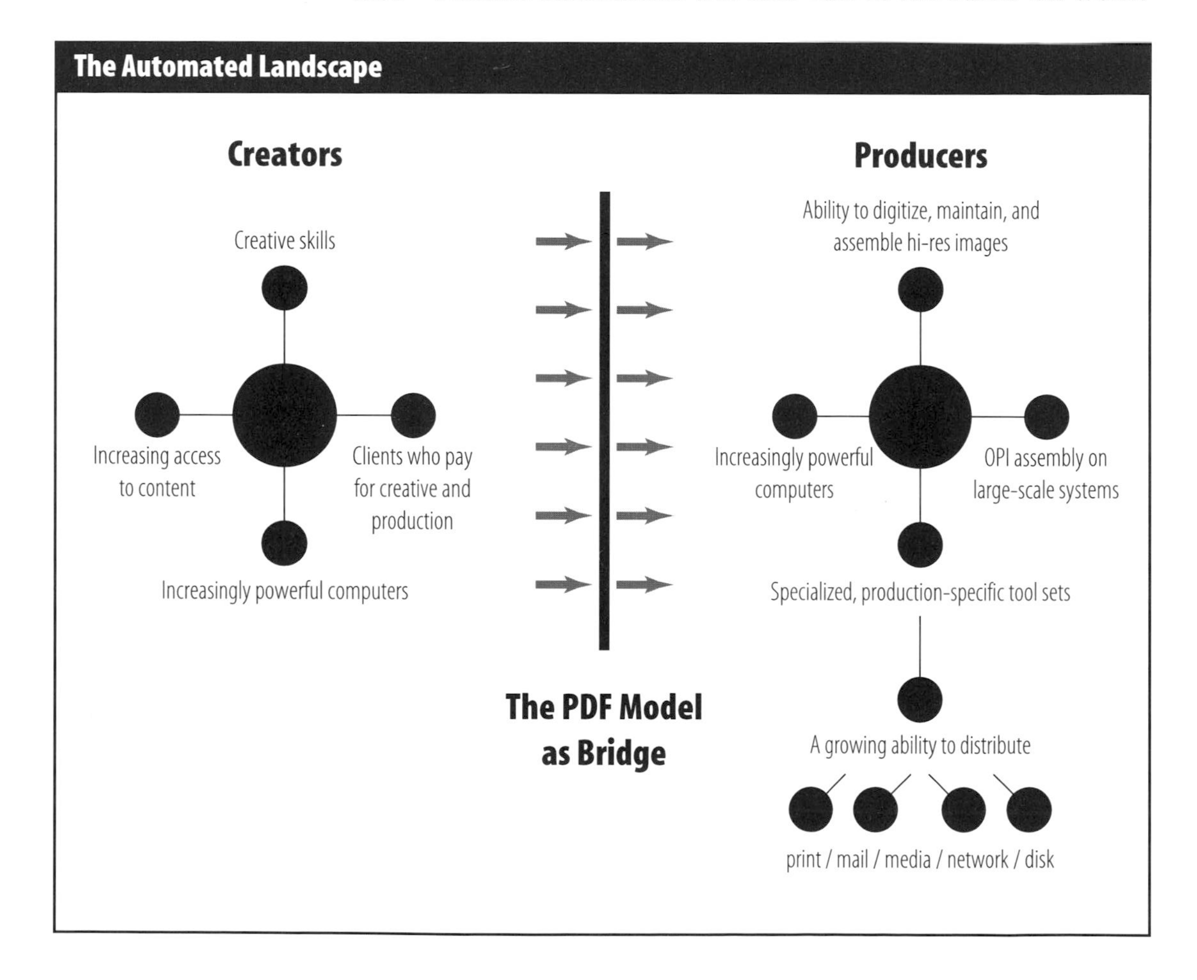

Requirements for an Automation Solution

The prepublishing industry needs a solution that runs on standard platforms, currently Macintosh and Windows. (Like it or not, UNIX is not a standard platform in the graphic arts industry.) Although many trade shops have, in fact, moved to UNIX servers in an attempt to solve throughput problems, having to deal with UNIX in a typical mixed environment, where you have Windows and Macintoshes running together, comes with a very high maintenance cost. You really have to know UNIX to process Macintosh or Windows pages through UNIX platforms. Even though the UNIX platform provides advantages in raw horsepower, its complexity offsets any productivity gains. Beyond that, Macintosh and Windows platforms are starting to show rapid increases in processing and input/ouput speeds, increases that are making both platforms considerably more useful for intensive operations.

The industry also needs a solution that provides control. A solid workflow-control product should operate very similar to the programs and events over which it provides the desired control. In short, you shouldn't need a degree in programming to figure your way through the design of an automated event sequence. It should be a relatively natural function—related to the other functions you normally perform. If you have to move from a Mac to an Intel box running Windows, you don't get the feeling (anymore, and probably to a greater degree as time goes on) that you're moving from Ireland to Ecuador. The climate, which can be thought of as how a particular program or OS expects you to respond, isn't as different as it once was.

Finally, our industry needs a stand-alone application that controls other applications. Many attempts at workflow-based layout and manufacturing solutions have had a good portion of the control code built directly into the applications themselves. A development team working on a photo-editing program typically won't spend time writing scheduling, or accounting, or workflow control codes into that application. The most they might do is make certain data sets readable by the workflow tools a shop might be using. Things like preferences files and other inherent controls should be able to provide data streams transparently of the core, native functions built into the software.

This desire to write nonrelated code plagues our industry today. We strongly believe that in order to meet the widely divergent needs of the different environments in which it might be used, software should evolve to be much more modular in nature. A functional workflow/control package would be highly modular. That isn't the case with most packages currently available. (This is another wrinkle in the UNIX/minicomputer approach to workflow design. Although the multitasking features are highly important, it's very, very difficult to anticipate one operating system's nuances while programming in another.)

There are basically three different levels at which a workflow product might read or act upon the data provided by the applications:

- At the first level, the creation or production software might be highly aware of the workflow package, and literally be "talking" with it while starting, processing, and completing specific functions within the event sequence.
- At the second level, the program might be partially aware of the workflow-control program, and be able to send, but not receive, communications to the control.
- At the third level, a program might know nothing at all about the fact that it was triggered (opened, invoked, run, or even controlled) by the control program. Increasingly, application programs are becoming scriptable, which would allow a control program to trigger such a third-level application without its direct knowledge. Naturally, a workflow that functioned with nothing but level 1 applications, which provides a high level of interapplication communication and pass-through variables, would be the most desirable.

Automation and the Role of Servers

Automation of any sequence of manufacturing events relies on a network making use of server technology. Within the context of our automation discussion, there are several types, or categories, of servers that we might encounter: the file server, the print server, and the process or workflow server.

File servers are already well-established as important components in today's creator and producer environments. Basically, the file server acts as a hub for project components contained in near-line or on-line work. (Recent server implementations allow for the retrieval of files throughout the network, so they may not be physically at the file server.) Off-line, or archived, materials usually do not occupy an active partition of the server, and at some sites, archive servers automatically backup jobs on one or more media based on programmed events (such as the generation of a proof), or at regular intervals.

The file server acts as a central exchange, where each workstation performing specific functions during a prepress task can look for and find the required elements. The use of a central file server reduces or eliminates many problems: duplicate versions of the same file, placing too high a demand on an individual station (centralizing storage requirements across the network by providing a central storage location), reducing processing logistics (in finding, saving, and moving files), and simplifying security and backup functions. In some cases, the file server might also provide computing functions (access to specialized applications) that are run from the server. The individual workstations, called *clients* (of the network), invoke the program that subsequently runs on the processor of the server, but sends results to the client workstation for further processing.

In many instances, specialized file servers provide not only the functionality that we just described, but add tool sets specific to the processing and management of high-resolution images.

An example would be Color Central, which provides the ability to control input, distribute images, generate OPI-referenced proxy images (FPO), local- and wide-area network distribution functions, and more. Fundamentally, however, an image server is basically a specialized file server, and ancillary tool sets only add value to the core functionality by providing centralized file storage and management. We have seen sites where the image server is run on a computer from the file server.

In the future, as output options increase, we will see an increase in the availability of automated tool sets that will provide solutions for cross-purposing.

The second type of server already found at many sites is the print server. A print server acts as a virtual output device; that is, it controls when and to which device a particular file is sent. In the design environment, the ability to free a workstation from the print queue is in itself a major productivity enhancement. In the production environment, multiple output devices are required: laser printers, proofing devices, imagesetters, and so on.

It's not at all uncommon to find three, four, or more imagesetters, for example, each capable of producing workable film under different conditions, all being at the end of the workflow sequence. Depending on the workflow and whether or not a particular device is available at the moment the file is ready, a print spooler or server is absolutely vital to effective operations—to an even greater degree than the freeing up of a designer's workstation. As effective as that might seem, print servers are the cornerstones of any automated production strategy. The more intelligent and powerful the print servers, the more automation a production site might hope to achieve.

The third type of server in our scenario is one that isn't commonly found in today's production environments: the process or workflow server.

As yet, the concept of a workflow server hasn't been discussed in any great length by our industry. Granted, design of and control over workflow sequences is beginning to be recognized as the next logical step in software tool development. In the next several years, we will see the emergence of process-control software begin to rival the current focus on file and color models. It is a very logical progression in the development of the digital communications industry. In a sense, we really can't move forward too much further without such development. As we've stated, processing digital content for distribution must eventually be based on digital, not conventional analog, processing strategies.

The OPEN Strategy

The basis of the OPEN strategy is quite simple: maximize any processes that can be executed without operator intervention. Such processes might be defined as any that could be triggered and then performed based on digitally available data coming from the actual content or content container being processed. The control software would be run from a server dedicated to that purpose. The controls thus provided would function equally well over local- or wide-area networks (LAN or WAN).

This control software would essentially provide seamless integration of functions normally available in two or more programs. The control software would, in essence, take the place of an operator who first has to open Application 1, perform a task, close the file, open Application 2, perform another task, and so on. Any sequence of events that might be defined by the team working on the project as being required could therefore be "programmed" into a simple event sequence called a "pipeline." This pipeline would represent an entire event sequence, and would be available either as a Chooser device or as a specialized, monitored "hot" folder. Once a job was sent to that Chooser device, or dropped into that special folder, the event sequence would begin, and sequential processes, executed from within disparate software applications, would be invoked and executed.

That, in a nutshell, is the OPEN strategy. It's elegant in its design and dramatically powerful in its potential. It finally provides automated process controls—to pave the paths that John Warnock described as having been blazed by early PostScript "explorers."

Today, OPEN runs on a Macintosh platform, but long before the concepts we're discussing in this book are out of date, it will be running on every major operating system.

In addition to its ability to seamlessly integrate the functionalities of different programs, OPEN provides additional tools to the workflow designer.

The first tool is the ability to control *messaging.* By providing constant feedback to management about the current state of any projects that might be living inside a pipeline, the software acts as a structured and highly accurate report generator.

Second, if a particular container runs into any problems during the execution of an event sequence, the software provides detailed *error logging* and allows for the movement of individual projects or components into predetermined areas, where they might be analyzed, corrected or completed, and sent back into the pipeline at the appropriate point.

The third tool OPEN provides the workflow manager is the opportunity to store or divert components into "buckets" where specialized, manual, or otherwise nonautomatable processes and events might be executed.

Gradual Automation

Like Rome, all roads in our industry lead to workflow automation. But also like Rome, that automation won't be built in a day. It is virtually impossible to imagine many chaotic sites trying to apply the OPEN concept to half-understood and chaotic workflows that many shops still labor under. Automating mixed-up workflows certainly won't lead to more efficiency—rather, to disaster.

But even the best-managed shops can't immediately apply the OPEN approach without careful study of existing workflows, as stated earlier in this book. And the first steps toward automation will have to be baby ones. It's unlikely that major operations can be strung together without serious inspection and trial at each stage.

But there are a number of steps that can be automated. As we have shown, the concatenation of steps that can be made into an automatic procedure must be done piecemeal at first, but the existence of a growing number of useful scripts can start the process. Most of these involve the moving, naming, and/or queuing of jobs between the completion of major operations (like color correction). It can also mean the updating of databases and tracking programs, as well as the conversion of file formats. One ideal to strive for is that an operator should never have any need, upon the completion of one job, to set about looking for the next one. It should be ready and waiting, with no further assembly required. Nor should any mad search through stacks of paper or labyrinths of magnetic media be required. Furthermore, the contractual requirements (color correction, trapping, proofing, imposition) should all be immediately available on the operator's workstation.

We recommend that after a solid graphing of workflows in your shop, you try to use available scripting tools to automate the most common and time-consuming procedures. Once those steps have been tested and debugged, incorporating them into an OPEN environment will be simple. And when your operators feel the incentive to eliminate excess steps and buy into the idea of continuous improvement, they will find ways to automate more and more processes.

CHAPTER 8

WHAT'S NEXT

It's clear that workflow is becoming more and more important to the graphic arts. Graphic arts businesses are getting increasingly worried about their continued existence. Producers are under growing pressure to reduce costs and turnaround time while increasing effectiveness. A number of new media, tied to traditional media by file formats and desktop computer platforms, are taking on new importance. In these new media, models for productive workflow are totally uncharted; distinctions between the creative and production functions are more confused; and the expectations for profitability and costs are still uncertain.

This is not to say that the workflow questions in the standard print media are by any means resolved. While we believe that the methods and strategies outlined in this book are good starting points, they are just starting points. From preflighting to data collection to automation of functions, the tools available now are just first generation.

A major advantage is that the desktop/PostScript revolution has moved most aspects of the creation and production cycles onto computers, mostly Macintosh computers. A majority of the work is based on PostScript to some degree, and there are standards developing for raster data, color synchronization, and fonts. Throughout this book we've noted the excess of options, but the overall trend is toward a discipline of the applications and formats that work.

This chapter assumes three constants for the near future in our industry: one is the PostScript language, the current center of the graphic arts industry; the second is the pre-eminance of Adobe's PDF as the portable graphics format between media and between platforms; and third, we see a product like OPEN (and others like Mainstream) behind the automation of document production. We believe that the dominance or eventual dominance of these products is the likeliest course. If PDF and OPEN are, in themselves, not dominant, we have little doubt that similar products will succeed and/or coexist.

The Basics

We believe there are three major areas where workflow enhancement is needed: at the system level, at the PostScript level, and at the automation level.

First, there is no fundamental, system-level strategy for facilitating the tracking and movement of digital pages through pre-planned workflow stages. Currently, the tools and hooks for doing this are few, largely custom-programmed, and hard to link to flexible report and data-control systems.

Second, although the ideal of PostScript is device-independence, in reality that ideal falls short. Decisions concerning final output conditions—which output device is best, what screen angles should be used, what trapping settings are best suited for that output device, which rendering (screening) methods are most suitable, and other issues—are being determined far too early in the creation process. This severely limits the choices available to a production site. For example, the issue of what color space to archive image files is starting to have serious implications. The concept of repurposing content for offset printing, for color-copier output, for a video presentation, and/or for CD-ROM means that the utmost flexibility in preserving graphic elements is starting to become a requirement for archiving.

And finally, although everyone seems to have computers and software to routinely perform creation and production tasks, there is very little automation evident in the typical workflow, even though automation has tremendous economic potential.

These three points are actually closely related, and they will have to be solved in the print media before they get applied to new areas. We do, however, believe that their importance cannot be overlooked.

Tracking and Routing Projects

The whole production system is a victim of what might be called "artificial stupidity," in contrast to artificial intelligence. While information vital to the production process begins to develop early on in the creative process (in digital form), there is no automated way to collect that data. Each native format stores document-specific information within its own internal file structures, but not in a way that can (easily) be collected from multiple, disparate applications. The information is there—it's just not being gathered. Native-format files are positioned on digital pages during the import event, containing far more information than is really necessary, yet ignoring or forgetting data that's very necessary later on in the process.

However, there are workflow permutations that can negate any advantage digital-document creation and production might offer. For instance, there can be too much rework, too many errors, too many people involved, and too much money being spent processing digital pages. Although the creative side of the equation has benefited greatly from the advent of desktop, the production community has not. Obviously, if production isn't running efficiently, creators are paying more and waiting longer for their work to be completed than they absolutely need to.

As we have stated elsewhere in this book, the collection of good data is a necessity. Ideally, all serious events should be reported to a master database. The status of each job and of each element in each page should be immediately available. Automated workflows should be generated, and a to-do list as well. All this should be done with minimal operator intervention.

The Role of PostScript

Although Adobe PostScript exists everywhere in the process, it wasn't designed to assist in the workflow process beyond its ability to describe resolution-dependent pages.

Since the PostScript model allows for such a broad definition of page content, it has proven almost impossible to stabilize in a given workflow, except in rare circumstances where the form of the document remains relatively constant from project to project, such as a periodical. Only mature periodicals based on proven templates seem to pass through the production process with no interference, and even then, the hiring of a new designer or production specialist can set the whole process back.

As creators and producers have come closer to the ideal of repeatability, many production sites around the country have gradually achieved a reasonable degree of economic success. Even then, however, many tasks are performed inefficiently. The power of PostScript-enabled creation tools allows such a degree of latitude in what might be found on a page or within a project, not to mention how it got there, that production facilities have had a very difficult time stabilizing their own process maps. Such plans work for a while, until a designer tries something new, or a new client is brought into the mix, or a new upgrade to a little-used software application is issued by a developer. Then a model developed just last week begins to break down.

In short, the strength of the PostScript model—its vast flexibility at the design stage—becomes a challenge when complex production tasks need to be carried out. Moreover, the freedom of PostScript constantly tempts ambitious designers to set their imaginations far beyond the limitations they encountered when pasting together a board mechanical. An increasing number of printed pages involve elements such as patterned page surfaces, rotated and distorted images, manhandled, shaped, or newly created type, complex gradients, and other elements far beyond the ability of any layout artist to create by hand. These tempting additions frequently present unrecognized production problems when output. This problem, however, is being addressed by Adobe's efforts with the Acrobat PDF model.

PDF and Document Portability

There exists a powerful economic stimulus to reduce the complexity of processing digital pages on both sides of the production equation. Too many people are currently involved in producing work from digital systems, and this labor overhead is directly responsible for costs being too high. The development of better interapplication communication, combined with a database approach to the elements contained in any project, will result in more efficient workflows. The industry will reduce its rework factor, which remains abnormally high. Many of the processes involved in digital workflows would become highly automatable if the proper tool sets existed. PDF, the Adobe Acrobat file format, seems to offer many of these features.

Document portability offers a fundamentally different approach to moving a project from conception to completion. For a document to be portable, it needs to be simplified. The term "distiller" accurately reflects the underlying strategy of the PDF model. The typical PostScript file can be complex and needs simplification—at the very least for specific stages of creation and production. When a producer site receives a digital file, the variables that can (and do) exist are overwhelming and are a hindrance to automation. Only through a reduction of these variables can we achieve true process automation for specific, well-defined workflows.

There will be an increasing demand for the creation of documents that are not—at least in the early stages of the creation process—destined solely for distribution as print media. Digital information is well-suited to alternative delivery modes, and the inclusion of non-print-specific components in the content list of many projects will grow to include motion, sound, and interactivity to meet these expanding delivery options. Document portability, to meet all possible conditions, must address non-print-specific components.

As workstations get cheaper, and tools begin to truly automate the collection and production processes, it will become easier for untrained individuals to publish their ideas. Distribution will no longer be limited to large runs, therefore making it economically feasible for more people to collect content in a digital form. In short, the pool from which digital documents originate will grow exponentially. There will be more creators and fewer—but far more efficient—producers. This is a reality that many producers are having a difficult time admitting to themselves.

The design process is well-suited to a high degree of device independence, and this trend will continue and be strengthened as the future unfolds. PDF architecture will need to expand its ability to communicate critical production-related information (such as OPI referencing, standard color elements, or spot colors) early on—and in a highly transparent manner—in this process of portable document creation. Few, if any, decisions about production-related issues should have to be made by the designer—other than those directly impacting final results, such as stock, specialty inks, trim sizes, bindery options, and similar considerations.

The resultant creative effort will be delivered to production professionals. Despite the fact that some blurring of responsibilities between creator and producer has occurred, this trend will reverse itself as more production-specific tool sets become available. The current trend is a direct result of tool-set limitations—it is not, we feel, indicative of a desire on the part of the design community to assume responsibility for production functions over which they have little control and less desire to learn.

Device Independence

It will remain in the best interests of processing efficiency to delay the assignment of device specificity until the best possible time. By identifying those areas most suited to delayed device-independence—for instance, when to convert RGB to CMYK—the designer's tool sets should provide the producer with the information required to choose which final device or application is best applied to which process. Hence, the "database" approach to creator's software tools and the necessity for plug-in capability in production-specific tools.

Several key functions currently "straddling" the content/production environments are in need of a better workflow model. Among these are the handling of high-resolution color images, dealing with adjoining and/or overlapping objects of differing hues, page geometry, imposition, and raster-image processing. Efficient workflow reengineering demands that these functions, and possibly several others, be moved out of the creation cycle and delayed until the appropriate production event.

New Tools

Production facilities are in dire need of production-specific tool sets. Some examples already exist, as evidenced by products such as PressWise, TrapWise, OPEN, PrePrint Pro, and the Color Central OPI-referencing/server tools. This trend toward production-specific tools must continue and be strengthened. Software developers need to heighten their development efforts regarding the PDF model. A case in point would be a tool set that would allow preflighting of incoming digital mechanicals. Such a tool set would allow those individuals in positions of responsibility within the production environment to make time- and device-critical decisions about trapping, referencing, separation, and imposition at the correct time.

The economic and time-related motivation to simplify processing exists equally on both sides of the creator/producer equation. The designer would print the projects first to an in-house, early stage proofing device, and then to PDF files. Today's alternative—a complex and dedicated process of element collection prior to shipping the (overly large and complex) files to the producer—is imposed on the creative team. It's an extremely time-consuming process, rife with variability and the possibility of overlooking key components. If all PDF did was to simplify this process—which it already does to some extent—it would be enough to convince every designer to use it on every project they ever sent out.

In order to achieve the desired simplification of PostScript page processing, the PDF format (the structure of which is already extensible through plug-in technology) must be expanded to address several production-specific capabilities.

GLOBAL EDITING. The availability of a global editor capable of accessing individual page components originally created in a (known, or at least anticipated) native format. Such an editor would be capable of working on any document processed and distilled with a system-level driver such as the PDF Writer. This system-level application would, if necessary, be distributed free of charge to all creators. The editor would be capable of processing and expanding those transparent, production-specific function sets contained in the creator application, such as OPI referencing, imposition issues, geometry, etc.

BETTER COMMUNICATION TOOLS. Improved tools to facilitate interapplication communications between open-architecture systems (the Macintosh/Intel environment) and proprietary systems such as Hell, Scitex, Crosfield, and others. A gap still exists—and will probably continue to exist for marketing and political reasons not related to processing efficiency—between native formats in the two environments. It cannot be assumed that companies like Scitex will abandon key strategic technologies, such as the Scitex CT-Linework model found in its native firmware and software architecture. Therefore, it will be necessary to create and increasingly rely upon conversion plug-ins that will facilitate content movement across such disparate platforms.

CALIBRATION SUPPORT. A plug-in capable of calibrating base-page color elements (such as those created within the content of a page-layout program) with elements that will, during the production process, be integrated into the imposition at some stage prior to the file being sent to the marking engine. This tool would be able to reference the applicable color profiles utilized by the creative team against those applicable to the specific input device/process available at the production site.

STORING HI-RES ELEMENTS. A plug-in somewhat related to calibration would allow the high-resolution elements to be input and directed to the proper (under flexible and changing availability conditions) storage location within the production environment. In an alternative scenario, this plug-in would allow for the identification and classification of preexisting elements, which may be either off-line or near-line at that specific production site. This is an important consideration that optimizes the process of image input. Since so much of the publishing process is accomplished through the use of pick-up components, the plug-in should be able to either a) generate a new input or an order for the task to be performed at some parallel point, or b) automatically identify, locate, and generate the required OPI referencing instructions. At this point, such contone elements would be calibrated in a full-gamut model, such as RGB or CIE. Device-specific instructions (such as the conversion to the highly specific CMYK model) would therefore be delayed until later in the process. All of this information would be input into the system at various stages of the creative and production process.

JUST-IN-TIME CHANGES. A plug-in that would allow specific operators access to high-resolution or prerendered low-resolution images contained in the incoming PDF file. Depending on the exact workflow, this plug-in would allow changes to densities, color casts, and screening, and would allow the scheduling of events such as screening, trapping, and rendering.

In general, our industry needs the comprehensive integration of widely disparate software tools to address the realities of creator/producer strategies, as well as the producer's need to conserve resources. It's futile to attempt to solve the broad problem of document complexity and inflexibility by having everyone use the same tools to accomplish widely divergent functions. What's needed is an underlying—and again, largely transparent—software structure that frees the creative team from having to make production decisions that are best left to experts. Until new tools and PDF plug-ins address these issues and allow the creative teams their inherent right to the creative process while providing the maximum, device-specific flexibility to the production group, production sites will continue to operate in a world dominated by costly equipment, happenstance workflow models, and expensive, manufacturing-savvy consultants.

Additionally, and in parallel, creative and production-specific tools must expand to include job components not related to print-only production. The integration into the workflow process of such elements as video, animation, audio, and user interactivity will further challenge existing workflows and create an even greater need to simplify—and keep simplified—digital documents at various stages during their creation and subsequent production.

When to Do What?

The order in which production-specific functions are applied to Postscript/PDF documents will be improved.

Imposition, which is obviously the highest manifestation of device specificity, is currently found at the end of the production cycle, with individual page-form processing being the focal point of most existing workflows. This is the case throughout the industry. Key elements of imposition and final assembly should occur prior to many of the functions now performed before imposition. In particular, the functions of OPI referencing, separation (perhaps even size- and resolution-specific scanning) of continuous-tone color elements, and trapping are currently assigned and executed too early in existing workflows.

The basic premise is really very simple: output devices are a lot faster than current workflows would indicate. The actual marking engines are often waiting for processing-intensive tasks that could be moved upstream, toward the conjunction of the creation and production processes.

OPI REFERENCING. The connection between low-resolution (FPO) images and high-resolution, size-specific images, which are necessarily much larger, should occur before the imaging process begins at the marking engine. If imposition rules are applied prior to the referencing process, the result is a much more efficient imposition. When digital composition of high-resolution images occurs, it should be done on preimposed film. This satisfies the need to reduce the movement of large files, and allows the imposition process to happen on a much smaller, and hence much more efficient, document. The disk space, time, and processing power required to generate digitally composed pages can be deferred until the last possible moment—immediately prior to RIPping and after the document has been imposed. This device-specific event has no impact on the sequence, or final results, of the OPI-referencing and composition function. If OPI happens at the time the file is RIPping, and prior to the creation of press-ready signatures, it delays the process and requires that imposition instructions be performed on huge files. The increase in efficiency gained by preimposing flats prior to the execution of OPI referencing will result in dramatically improved workflows. This scenario requires that imposition instructions be made part of the PDF specification set.

RGB/CMYK CONVERSIONS. Separations should also be done after page imposition, for many of the same reasons. Separation and OPI referencing should be made part of the same postimposition sequence. Elements such as color bars, screen/density/coverage targets, and similar components should be put into all files by the creation tools, eliminating the need for production personnel and resources to focus on elements that remain device-independent throughout the workflow. In this workflow, the interpretation of all contained color elements could be done in one pass, as opposed to the current model, which has each color element processed at least twice (or more): once individually and then again as part of a massive RIP-time event.

TRAPPING. Intimately related to OPI referencing, trapping is also best performed after the imposition event. In a perfect world (which is what we're trying to describe here), the best trapping would result from a postimposition effort. In that manner, actual press mechanics (gripper position, running direction, ink-to-ink press positioning, more accurate stretch factors, etc.) would be figured into the trapping equation, resulting in far more accurate final results. Electronics should produce better results than conventional processes—a goal that, although in sight, is not clearly visible under current operating procedures.

RENDERING. The actual image-screening process, which by definition is the most device-specific function in the entire process, should be performed as a final rendering process, with decisions about the exact nature of the screens employed (that is, conventional, stochastic, FM, etc.) being delayed until the last possible moment—again, to meet the fluid conditions needed to achieve the required flexibility of on-the-spot device assignation within the production environment.

Can You Automate a PostScript Page?

In our discussions with experts in the industry, it was often stated that there is so much variability in how a PostScript page might be processed that any attempts at automation are doomed from the start. We strongly disagree with this statement, for a variety of reasons.

First, this isn't brain surgery. Granted, on first inspection one might agree that the task is Herculean; there is a tremendous amount of variability in file structures. But despite this, there are sites such as Graphics Express, Lanman Lithotech, Gamma One, and thousands of others that successfully receive, output, and distribute PostScript pages. You can, with structure, training, and continuous refinement, make PostScript workflows work. Especially in the case of "repeatable" content types—ones where there is a great deal of structure to the creation process (for example, magazines, catalogs, annual reports, books, newspapers, packages, textiles, or interactive development)—we know what must be done to incoming files to achieve the desired results. It's done every day. If it can be done by six different people working in concert at the production site, it can probably be done with six programmed events. This may seem harsh, and clearly points to further attrition within the highly skilled arena of production, but it is also very true. Processes can be automated with a sufficiently powerful and flexible control package like OPEN.

A second factor that we disagree with is that attempts at automation are doomed from the start. Variability will always exist. When you're seeking standardization or stabilization of a process or an event sequence, you begin by looking for *commonalities,* not the *differences.* In other words, you automate what can be automated, and refine the effort from there. You don't sit around and try to figure out what rare and occasional circumstance would stop the process. Let's say a production site finds that 70% of its work is catalog or magazine (or any other repeatable, structured document) work, and 30% comes from the city's top agencies. Clearly, the agency work is far more flexible in nature and less likely to benefit as much from an automation effort than the repeatable work. But that is not justification for failing to automate what can be automated. This sounds a little foolish: why would anyone not automate the magazine work because the same strategy wouldn't benefit the agency work to the same degree? Look around and you'll see that mentality at work everywhere. We remember, in the early days of desktop, hearing, "What about my existing flats? How can I turn them into digital information? If I can't, what sense is there in buying into desktop?" The obvious flaw in this logic is that there wasn't much to be gained by digitizing existing flats—but buying into desktop maximized the most important function: satisfying an increasingly computer-literate client base. The same logic holds true here. As software becomes more and more modular in nature, what it does will increasingly fall into automatable event sequences. You can't afford to wait until everything in the world can be controlled by something like OPEN. You need to do it now, and learn as its capabilities expand based on supporting development in creator and manufacturing-specific software. The writing is bold on the wall: in the production shop of the future, control software will be opening, saving, executing, and triggering events—not an assembly line of highly skilled individuals. We've said it before and it bears saying again: the paradigm of our industry is changing from a skills-based, event-oriented industrial model to a digital, automated production model: *automated PostScript production.* Again, since Adobe owns PostScript, and now PDF, they are likely to evolve into the dominant player in process-control software. Adobe knows the content better than anyone, and possesses incredible programming resources.

Conclusion

While we don't believe that anyone can predict with accuracy the state of the graphic arts industry in any time span greater than a few years, and some would argue months, we predict that the following key trends will emerge.

The fact that digital content is dominating the work profile coming into production facilities calls for digital workflow solutions. In large part, the problems we are now encountering are best approached by digital, not analog, production strategies.

The effort to automate will be motivated by increasing demands on production facilities to reduce costs, increase turnaround speed, and provide a high level of flexibility to a growing creator base working with widely disparate PostScript authoring tools.

Automation is very attainable. We know how to process PostScript pages. Even though there are still problems, file types and color models—the two primary obstacles in previous automation attempts—are rapidly stabilizing. This allows us to begin to develop automated workflows that depend on information available from increasingly stable applications.

Many operators currently required to perform manually invoked and executed digital events will increasingly find their responsibilities handled by process-control software.

And finally, shops that move quickly to automate those processes that can be currently automated will gain competitive and operational advantages over those that do not.

The challenge for any reader of this book is to reengineer current workflows so that they:

- Optimize the equipment and skills people have,
- Provide the maximum functionality to creators,
- Provide the maximum flexibility to producers,
- Delay the assignation of devices until the last possible moment,
- Reduce or eliminate errors as a project matures, and
- Automate the process wherever it make sense to do so.

Whether on-demand printing, video feeds, database publishing, direct-to-whatever, or any other technology dominates, the workflow questions will not change.

In closing, we would like to make a simple statement: technology will continue to change and products will be redefined. Profits will always remain what they are today—money above and beyond what it costs you to do business. If you focus on profits and the people you employ, before you spend your time floating around the floor of the latest trade show, you will probably find that success isn't nearly as elusive as it sometimes appears to be.

Bibliography

Burger, Rudolph. *Color Management Systems,* The Color Resource, San Francisco, 1993.

Crosby, Phillip B. *Quality Without Tears,* McGraw-Hill, Cambridge, Mass., 1984.

Davidow, William H., and Michael S. Malone. *The Virtual Corporation,* Harper Business, New York, 1992.

Field, Gary G. *Color and its Reproduction,* GATF, Pittsburgh, 1988.

Hammer, Michael, and James Champy. *Reengineering the Corporation,* Harper Business, New York, 1993.

Katzenbach, Jon, and Douglas Smith. *The Wisdom of Teams,* Harper Business, New York, 1993.

Kieran, Michael. *Understanding Desktop Color,* Peachpit Press, Berkeley, California, 1994.

Lawler, Brian. *The Color Resource Complete Guide to Trapping,* The Color Resource, San Francisco, 1993.

Pall, Gabriel A. *Quality Process Management,* Prentice Hall, New York, 1987.

Southworth, Miles, Thad McIlroy, and Donna Southworth. *The Complete Color Glossary,* Southworth, McIlroy and Southworth, Rochester, New York, 1992.

INDEX

Steve Hannaford is a writer, speaker, and consultant on prepress issues. He has been writing and consulting about desktop color and publishing since the first color separations were tried on the Macintosh. Hannaford is editor of *Prepress Business Observer,* a newsletter about the business side of the industry. He is also editor for the *IDIA* (formerly *AISB*) newsletter. He is the author of Agfa's *Introduction to Digital Color PrePress, Volume 1,* and *The White Paper on Desktop Publishing,* and a contributing editor to *MacWeek* and *Step-by-Step Electronic Design*. He has written for a wide variety of magazines, ranging from *Publish* to *Pre-* to *Digital Imaging,* and has spoken at many conferences and shows, ranging from Graph Expo to Conceppts to Seybold. He has consulted for a variety of service bureaus and prepress shops, along with several national equipment vendors. His background is in computer graphics and computer systems, and he has been a publisher for the last six years. Hannaford received a Ph.D. from the University of Toronto.

Gary Poyssick is an industry consultant specializing in design and workflow management training that focuses on better profits, productivity, and rework reduction. Gary is the author of *A Printer's Guide to Desktop Publishing, The Electronic Cookbook, Creative Techniques: Adobe Illustrator* and *Creative Techniques: Adobe Photoshop,* as well as a suite of training products for the professional graphic arts community. His consulting clients include: World Color; 3M Corporation; *Smithsonian Air & Space* magazine; Apple Computer; General Mills; The Lanman Companies; *New England Journal of Medicine;* Cadmus Corporation; Time/Life Books; Publishers Clearinghouse; and many others. Gary currently serves as product development manager for Against the Clock, a nationally recognized firm that develops, designs, and markets workflow-based curriculum materials for creator and producer environments (http://www.atclock.com). Gary lives in Tampa, Florida, and can be reached via e-mail on the Internet at courseware@interramp.com.

Adobe Press books examine the art and technology of digital communications. Published by Macmillan Computer Publishing USA, Adobe Press books are available wherever books about computers or the graphic arts are sold.